HAM RADIO
FOR BEGINNERS

The Complete Guide to Mastering Your Radio from Everyday Use to Emergency Situations. Get Ready for the Technician Class Amateur Radio Exam

By
Everyday Expert

© **Copyright 2024 Everyday Expert - All rights reserved.**

The contents of this book may not be reproduced, duplicated, or transmitted without the direct written permission of the author or publisher.

Under no circumstances will the publisher or author be held liable for any damages, recovery, or financial loss due to the information contained in this book. Neither directly nor indirectly.

Legal Notice:
This book is protected by copyright. This book is for personal use only. You may not modify, distribute, sell, use, quote, or paraphrase any part or content of this book without the permission of the author or publisher.

Disclaimer Notice:
Please note that the information contained in this document is for educational and entertainment purposes only. Every effort has been made to present accurate, current, reliable, and complete information. No warranties of any kind are stated or implied. The reader acknowledges that the author is not offering legal, financial, medical, or professional advice. The contents of this book have been taken from various sources. Please consult a licensed professional before attempting any of the techniques described in this book.

By reading this document, the reader agrees that under no circumstances will the author be liable for any direct or indirect loss arising from the use of the information contained in this document, including but not limited to - errors, omissions, or inaccuracies.

Table of Contents

Chapter 1: Introduction to Ham Radio 6
What is Ham Radio? 7
Brief history of Ham Radio 8
Importance of Ham Radio in Today's World 9

Chapter 2: Understanding the Basics 11
Fundamentals Concepts of Radio Communication 12
Frequency Bands Explained, HF, VHF, UHF 13
Modes of Communication: Voice, Morse Code, and Digital 15

Chapter 3: Getting Started With Ham Radio 17
The Role of Amateur Radio Operators 18
Equipment Overview: Transceivers, Antennas, and Power Sources 20
Setting Up Your First Station 21

Chapter 4: The Ham Radio Community 23
Joining Ham Radio Clubs 24
Making New Radio Friends 25
Participating in Contests and QSO Parties 27

Chapter 5: Licensing: Why and How 29
Understanding the Licensing Requirements 30
Types of Ham Radio Licenses 31
The Path to Getting Your License 33

Chapter 6: Preparing for the Exam 35
Study Resources and Tips 36
Understanding the Question Pool 37
Practice Exams and Effective Study Strategies 38
Exam Day: What to Expect 39
HERE IS YOU FREE GIFT! 41

Chapter 7: Operating Your Ham Radio 42
Making your First Contact 43
Operating Procedures and Etiquette 44
Using Repeaters for Extended Range 45

Chapter 8: Exploring Advanced Topics 48
Digital Modes and Software-Defined Radio 49

Satellite Communications and EME (Moonbounce) 51
DXing and Awards 52

Chapter 9: Ham Radio for Emergency Communication 54
Role of Ham Radio in Disasters 55
Getting Involved with ARES and RACES 57
Setting Up an Emergency Communication Kit 58

Chapter 10: Building and Experimenting 61
DIY Projects for Ham Radio 63
Antenna Desing and Construction 65
Homebrewing Equipment and Modifications 66

Chapter 11: Continuing Your Ham Radio Journey 69
Advancing your Skills and Knowledge 71
Upgrading Your License 73
Staying Engaged with the Community 74

Appendix: Glossary of Ham Radio Terms 76

CHAPTER 1

Introduction to Ham Radio

Greetings from the wonderful world of Ham Radio, which you are about to embark on. This chapter serves as your gateway to an amazing global network that unites individuals from many backgrounds via the wonders of amateur radio transmission. Ham radio offers something for everyone, regardless of interests in technology, new hobbies, or unusual ways to socialize.

Imagine having a gadget in your hands and a doorway to connections and conversations with individuals you might not have otherwise met. You can communicate across continents, over the depths of space, and even bounce signals off the moon, all from the comforts of your own house! Ham radio aims to overcome gaps between people, cultures, and languages and convey communications.

Perhaps you're asking yourself, "What exactly is Ham Radio?" Fundamentally, amateur or ham radio uses assigned radio frequencies for non-commercial message transmission, wireless experimentation, emergency communication, and personal leisure. For those who operate amateur radio, the moniker "Ham" has a rich history that stretches back to the early days of radio.

However, Ham Radio is a community rather than only a technical pastime. This friendly, teaching-focused group is even more eager to learn. There is a space here for everyone, eighty years old or older. The appeal of Ham Radio lies in its capacity to evolve and progress alongside technology while preserving the human element of spoken communication. It's a hobby that mixes the excitement of technological advancement with the ageless attraction of human connection.

It may seem intimidating to begin your adventure into Ham Radio, but remember that everyone starts out as a beginner. To enjoy Ham Radio, you don't have to be an engineer or scientist. All you need is a spirit of adventure, curiosity, and openness to learning. The group can provide advice, information, and the kind of companionship that results from similar experiences and passions.

As you delve into the world of Ham Radio, you will come across a diverse range of things. Along with comprehending the essentials of radio theory and electronics, you will discover how to establish your initial radio station and initiate your first contact. These initial steps can be exhilarating as they open up countless possibilities to acquire knowledge from each interaction.

Respect and safety are the topmost priorities in the Ham Radio community. As you gain more knowledge and proficiency, you will realize the importance of following the guidelines set by your nation's communication authorities. These rules ensure all users can use and access the airwaves without issues.

As we proceed through this book, we'll delve more into the facets of Ham Radio that make it such a fulfilling activity. You're in for many experiences, ranging from the technical aspects of setting up your station to the excitement of competing in events and contests. During our discussion, we will highlight the crucial role played by Ham Radio in emergency communication, as it acts as a backup in case traditional communication networks fail.

On this adventure, you'll learn that Ham Radio is about the people you meet along the road, not just the gear and signals. It's about creating a worldwide network of enthusiasts who share your interest through story-telling, idea-sharing, and community-building.

Thus, inhale deeply and advance with enthusiasm and an open mind. The world of amateur radio is prepared to embrace you. This is where your journey begins, and who knows where it will lead? The possibilities are as endless as the airwaves themselves.

Greetings from Ham Radio for Novices. This begins your journey into the exciting realm of amateur radio communication. Let's set out on this trip to explore, learn, and connect in ways you never would have thought possible.

What is Ham Radio?

You may ask yourself, "What exactly is Ham Radio?" as you set out on this adventure into the world of amateur radio. Together, let's set out on a journey to explore the spirit of this fascinating pastime that unites people worldwide.

Fundamentally, amateur radio, or ham radio, is a communication method that connects people over radio frequencies. It's a pastime beyond just conversing; it involves discovery, education, and even lending a hand when someone is in need. Imagine communicating vitally during an emergency when all other systems are down or sending a greeting to someone in a far-off country without requiring the internet or a cell network. That is the essence and strength of Ham Radio.

Ham Radio is your pass to a world stage where the participants are as varied as the discussions they have. The Ham Radio community is a melting pot of ideas, cultures, and friendships, encompassing young learners, seasoned professionals, scientists, and artists. There, those willing to share and learn take to the airwaves as their curiosity propels them on their journey of discovery.

A background in electronics or in-depth technical knowledge is necessary to start with Ham Radio. Although those abilities can improve your experience, they are optional to join this active community. Fundamentally, the joy of discovery and a want to communicate are at the heart of Ham Radio. The community can help and advise you, whether you're experimenting with antennas or making your first nervous broadcast.

One of Ham Radio's most lovely features is its accessibility. To get started on your expedition, you only need a radio, an antenna, and a transmit license. But the pastime's complexity is hidden by its simplicity. As you develop, you'll come across routes that lead to satellite communication, digital modes, and even moonbounce, the phenomenon of messages rebounding off the lunar surface. There are countless options for exploration within the vast and profound field of ham radio.

However, ham radio is more than a technical pastime—it may save lives during emergencies. Ham radio operators are essential conduits between affected communities and emergency services during natural disasters when traditional communication networks are disrupted. This facet of Ham Radio personifies the community and service mentality fundamental to the hobby. It's about changing the world with your enthusiasm and abilities.

Participating in ham radio also entails leaving a legacy. The history of ham radio tells a captivating story of invention, exploration, and the unwavering human need to connect. When you join our community, you're not just taking up a pastime—you're immersed in a narrative that spans more than a century.

Remember that every Ham Radio operator's path differs as you embark on this adventure. Some enjoy developing their own equipment, some enjoy chasing faraway signals, and others enjoy the companionship of local groups and competitions. Your hobbies, your curiosity, and the relationships you make along the road will all influence your journey.

Finally, Ham Radio is a call to inquiry, involvement, and connection. It's about extending your reach over the airwaves and unlocking endless opportunities. Ham radio offers a rich and fulfilling experience, whether you're drawn to the technical difficulties, the excitement of competition, or the straightforward pleasure of communication.

So, start out with an open mind and an inquisitive heart. Amateur radio offers a world of possible friends, adventures, and knowledge. Welcome to a pastime that's about more than simply radio waves—it's about connecting people and eradicating barriers. Greetings from Ham Radio.

Brief history of Ham Radio

Take a historical tour to discover the origins of Ham Radio, a pastime that has influenced society, invention, and history. Learning about amateur radio involves more than simply comprehending the here and now; it also involves appreciating the diverse range of narratives, innovations, and fervent personalities that have influenced its development.

Envision the late 1800s, an era of innovative innovations and unending curiosity. Heinrich Hertz's discovery of radio waves opened up new possibilities, and before long, explorers like Guglielmo Marconi were transmitting signals over the Atlantic. Radio communication began during this time, laying the groundwork for later developments into Ham radio. These pioneering explorers were the first to set foot on the airwaves, which were a huge, unexplored ocean at the time.

Amateur radio operators, also called "hams," started to appear as the 20th century progressed. These folks experimented with improvised equipment in their basements and backyards, captivated by the possibilities presented by radio waves. Despite not having any formal training in radio, their enthusiasm and creativity allowed them to accomplish amazing communication achievements.

The word "ham" itself dates back to the early years of the hobby. Although there are many theories regarding its origin, a widely accepted one holds that it originated from the disparaging epithet "ham-fisted" operators, which professional telegraphers used to characterize amateur operators who lacked ability. The term "ham" became proudly associated with the amateur radio community throughout time, representing their adventurous and can-do attitude.

The earliest amateur radio clubs and organizations were founded in the early 20th century, and they were instrumental in advancing the pastime, disseminating information, and standing up for the rights of amateur radio operators. These groups, which included the American Radio Relay League (ARRL), which was established in 1914, grew to be important hubs for the Ham radio community and helped to create a supportive and friendly environment that endures today.

The history of Ham Radio's significance in times of crisis is among its most interesting features. Amateur radio operators were asked to aid their nations during both World Wars, use their expertise to maintain contact and get intelligence. Ham operators have often stepped in to offer vital communication between impacted areas and emergency personnel during natural catastrophes when conventional communication systems frequently fail. The Ham Radio ethos is built on this heritage of service, which shows how the hobby has far-reaching effects beyond simple enjoyment.

Ham radio underwent tremendous transformation in the 20th and 21st centuries due to new opportunities for invention and communication, such as the Internet, satellite communication, and the introduction of digital modes. However, Ham Radio's core is still the same—a group of enthusiasts committed to learning more about the potential of radio communication.

As you prepare to embark on your ham radio adventure, recall that you are entering a tradition of trailblazers, innovators, and public servants. The spirit of Ham Radio is a monument to human curiosity and connection, from the early experimenters who were amazed by the power of radio waves to the current operators who push the frontiers of technology and service.

This succinct history serves as both a guide for the future and a look back. It reminds us that amateur radio is a heritage rather than just a pastime. When you turn on your radio and establish your first contact, realize that you are joining a chorus of creativity, discovery, and community that has been around for more than a century—a chorus that extends a warm welcome to you.

Importance of Ham Radio in Today's World

In a world where digital communication is ubiquitous, one may question the continued relevance of amateur radio. With social media, instant messaging, and cell phones at our fingertips, communicating via radio waves may seem archaic or even out of touch. However, one cannot emphasize the significance of Ham Radio in the modern world. It serves as a link between the past and the present and embodies the values of creativity, community service, and human connection.

Imagine a world in which the modern comforts of communication that we take for granted—like cell towers and the internet—suddenly become inaccessible. It's in these moments that the real worth of Ham Radio becomes apparent. Due to its independence from commercial networks, the ham radio is a vital tool for emergency communication. Ham operators are crucial in facilitating rescue and relief efforts during natural disasters by acting as unsung heroes.

However, the importance of Ham Radio goes beyond its use in emergencies. It's a testing ground for new and inventive technology. A radio can be a tool for learning and exploration when it is in your hands. You'll investigate the complexities of electronics, delve into the secrets of radio waves, and play around with digital communication methods. This pastime pushes you to solve puzzles, think critically, and learn new things. It's an exciting method to encourage young brains and seasoned professionals to push the frontiers of what's possible and cultivate a lifelong interest in STEM (Science, Technology, Engineering, and Mathematics) sectors.

In addition, there is a thriving global network of hobbyists involved in Ham Radio. Since it cuts over boundaries of geography, politics, and culture, it makes real connections possible between people who might never cross paths in any other setting. In today's fractured digital landscape, these connections are based on mutual respect and a shared enthusiasm for communication—qualities that are becoming increasingly common. Connecting with other hams will enhance your life in unanticipated ways as you will share more than just call signs; you will also trade tales, experiences, and thoughts.

Ham radio also provides a rare chance for introspection and mindfulness in today's hectic society. Patience and attention are needed when tuning into the airways, changing antennas, and searching for signals. It's an opportunity to take it slowly and pay close attention to the environment around you and the voices and codes that crackle through the ether. This part of the pastime fosters a sense of presence frequently lost in our daily hustle, encouraging a better appreciation for the straightforward yet profound act of listening.

Additionally, Ham radio is essential to scientific study and education. Amateurs advance our knowledge of space exploration and the Earth's atmosphere through endeavors like satellite operations and moon bounce communications. Education institutions worldwide integrate Ham Radio into their curricula, employing it as an interactive instrument to instruct students in physics, electronics, and communication theory. The hobby's significance in motivating the upcoming generation of scientists, engineers, and inventors is highlighted by its educational value.

Lastly, the spirit of exploration is personified by Ham Radio. It challenges you to set out on a voyage of adventure in which each turn of the dial unlocks a new area for investigation. Ham radio makes communication an adventure, reminding us of the excitement of the unknown and the delight of discovery, whether conversing with someone on a foreign continent, bouncing signals off the ionosphere, or making contact with a lonely island.

In the modern world, Ham radio's significance is vast and varied. By providing a forum for technical innovation, global community, mindfulness, education, and adventure, it closes the gap between technology and mankind. Recall that as you set out on this adventure, you're not merely investigating a pastime but becoming part of a worldwide community that prioritizes service, education, and connection. Greetings from the world of amateur radio, where the possibilities are endless, and the journey is only getting started.

CHAPTER 2

Understanding the Basics

Greetings, and welcome to Chapter 2, where we will gently delve into the principles of Ham Radio. This adventure is about setting the foundation for your future as a skilled amateur radio operator. Let's approach these fundamentals with an open mind and curiosity, remembering that every expert was once in your position.

Let's start by discussing the components of a Ham radio setup. Fundamentally, you will require an antenna, a power supply, a transmitter, and a receiver (frequently combined to form a transceiver). Think of these as the components of a recipe; each is essential to the result but requires some knowledge to use well.

The receiver records broadcasts from other operators, while the transmitter emits digital signals, Morse code, or your voice across the airways. By combining the two, the transceiver provides efficiency and ease. Imagine yourself having a discussion using the same device used for both listening and speaking.

The antenna serves as a link between your device and the wide spectrum of radio waves. It's more than simply a metal component; it's the lifeline of your signal, extending into space. Your ability to communicate over long distances might be significantly impacted by the kind and location of your antenna. See it as an amplifier for your voice, spreading the word about what you have to say.

Power sources come in many forms, such as solar panels, batteries, or plug-in energy. This adaptability guarantees that you may use your radio in various situations, enabling communication in isolated areas or during blackouts. It's the beating heart of your setup, giving your radio the energy it needs to function.

Let's investigate the frequencies now. A vast spectrum of frequencies is used in ham radio operations, and each has special features and rules. These frequencies offer unique opportunities and problems, just like different TV stations. While some can go across countries or seas, others can reverberate off the ionosphere and be perfect for local talks. Knowing these frequencies can help you communicate securely and successfully by directing you where and how to go. It's like understanding the rules of the road.

Licenses are another important factor. You must obtain a license from your nation's communications authority to operate a Ham radio. Typically, this process includes completing an exam covering technical knowledge, operating procedures, and fundamental legislation. Consider it similar to obtaining a driver's license but for the radio. It's a milestone that creates a plethora of opportunities for communication.

Remember the value of protocol and decorum as you embark on this adventure. Transmitting is only one aspect of ham radio; another is being a responsible and respectful community member. Your experience will be more pleasurable, and you will get respect if you can learn to listen before you transmit, use call signs correctly, and obey the existing regulations.

Making your first contact, or "QSO," is one of the most exciting aspects of starting with Ham radio. This is a period of great excitement and anticipation as you formally join the worldwide amateur radio operator community. Every QSO, whether a local discussion or a worldwide exchange, is a step toward boosting your self-esteem and growing your social circle.

Finally, keep in mind that patience is your friend. Learning Ham Radio requires time and practice, like learning any new skill. There may be times when signals seem hard to follow if there are technological difficulties. Nevertheless, you'll discover you're more capable and fully involved with every challenge you overcome. Honor your accomplishments, accept lessons from your errors, and value the trip as much as the final destination.

We have established the groundwork by reviewing Ham Radio's fundamentals in this chapter. Remember that every dial turn, transmission, and new information you pick up will help you progress from being an operator to a large, active community member. Greetings from the world of amateur radio, where every voice counts, and every link opens up new possibilities.

Fundamentals Concepts of Radio Communication

You're about to embark on an exciting and educational journey into the world of Ham Radio. To get the most out of this trip, you must understand the basics of radio communication. Let's dissect it using straightforward language so you may comprehend the fundamental ideas behind the radio talks you have.

The magic of radio waves lies at the core of ham radio. Consider tossing a stone into a still pool. The outer ripples resemble radio waves passing through the atmosphere. These waves transmit digital data, Morse code, and your voice over short and long distances, enabling communication.

Electromagnetic energy takes the form of radio waves, which resemble visible light but exist at invisible frequencies to the human eye. The frequency, expressed in Hertz (Hz), determines the precise channel your radio uses to send and receive messages and the pitch of the sound you hear in the audio. Similar to how you switch between stations on an FM radio, hams switch between frequencies to locate data, conversations, or broadcasts regarding emergencies.

The frequency spectrum is one basic idea you'll come upon. Imagine it as a huge ocean with distinct zones set aside for different forms of communication, such as amateur radio, television, aircraft, and maritime. Ham radio operators are allotted certain "bands," or ranges of frequencies, to operate within this enormous ocean. Every band has unique properties that make it appropriate for various forms of communication, times of day, or atmospheric circumstances.

It's important to know how radio waves move. They can go directly from one antenna to another, reflect off mountains or buildings, or bounce off the ionosphere, a layer of the Earth's atmosphere energized by the sun. This adaptability enables a variety of communication techniques, such as speaking with someone in your neighborhood, sending messages across continents, or even sending signals back and forth from the moon!

The mainstays of your radio station are the antennas. They are the devices that convert electrical energy from your radio into radio waves and vice versa; they are not just pieces of metal. Antennas come in various varieties, each having advantages and uses. Certain models excel in detecting far-off signals, while others are engineered to project your voice across vast distances. One of the pastime's most interesting aspects is learning about antennas and how to best utilize them for your purposes.

An additional crucial idea is modulation. It's the process by which a radio wave is mixed with your information—voice, Morse code, or digital data—to enable transmission. There are other techniques, but two that you'll likely hear about frequently are frequency modulation (FM) and amplitude modulation (AM). FM modifies the frequency of the radio wave, whereas AM modifies the wave's power to convey information. Each has a purpose and set of qualities that affect your communication's scope, lucidity, and caliber.

In Ham Radio, a "radio station" can be as basic as a portable transceiver or as complex as a fully-equipped desktop setup with many antennas. The term "radio station" may conjure images of enormous towers and intricate equipment. Your exploration and communication center is your station. It's where you can learn about electronics and radio theory, play around with various gears, and, most importantly, connect with the world's Ham radio community.

Finally, the idea of licensing is essential. A license is required to operate on Ham Radio frequencies. The focus here is on ensuring that all operators comprehend the fundamentals of radio operation, safety, and etiquette, not gatekeeping. Preparing for and gaining your license is a significant accomplishment in your Ham Radio journey, providing access to a world of discovery, education, and friendship.

Gaining an understanding of these radio communication principles will not only help you with the technical parts of ham radio, but it will also introduce you to a community of people who share a passion for communicating with one another via airwaves, curiosity, and adventure. You're not just learning about radio with each topic you understand; you're also opening up new avenues for exploration, adventure, and connection in the exciting world of amateur radio. Greetings and welcome to your adventure into the world of Ham Radio, where the airwaves are alive with the possibility of lifelong learning, new experiences, and friendships.

Frequency Bands Explained, HF, VHF, UHF

Digging deeper into ham radio, you will encounter terminologies like HF, VHF, and UHF. Although these acronyms are merely additional technical terms to learn, they are essential for accessing the rich and varied world of radio communication. Together, we'll examine these frequency ranges and simplify their meanings so you can learn more about them and how they could affect your ham radio travels.

First, let's talk about the HF band, which spans 3 to 30 MHz. People frequently refer to it as the "shortwave" range. Think of HF as the frequency band's global traveler. HF signals can travel great distances, even worldwide, because of their capacity to reverberate off the ionosphere, a layer of the Earth's atmosphere. Because of this, HF is perfect for international communication since it allows you to communicate with individuals anywhere on Earth without requiring a satellite or internet connection.

Many classic Ham Radio activities, such as long-distance exchanges (DXing), contest participation, and cross-border emergency contact, occur on HF. HF's unpredictable nature and reliance on solar activity and atmospheric conditions give it its enchantment. You can communicate with someone on a different continent one day, then find it difficult to reach someone outside your nation the next. This fluctuation makes using the HF bands more exciting and challenging.

Going forward, VHF, or Very High Frequency, covers the 30 to 300 MHz range. VHF is especially helpful for local and regional communication; it's like having a trustworthy friend who is there for you any time. VHF waves follow a more direct route and are closer to the line of sight than HF transmissions. This indicates that although they typically don't bounce off the ionosphere, they can still travel great distances, particularly when using repeaters to increase the transmission range.

VHF is the way to go for regular contact, connecting to local networks, or participating in community activities. In addition, many public service communications, such as police, fire, and ambulance services, operate in this band; however, as a Ham operator, you will use different frequencies designated for amateur usage. VHF is ideal for local emergency communication because of its dependability and clarity, guaranteeing that you can provide assistance or remain informed when it matters most.

Lastly, there is Ultra High Frequency, or UHF, which ranges from 300 MHz to 3 GHz. Consider UHF as the urban adventurer if HF is the globetrotter and VHF is the trustworthy friend. UHF frequencies are perfect for communication in cities where buildings and other structures can block signals because of their great penetration of urban surroundings. Like VHF, UHF mostly functions in line-of-sight; however, because of its shorter wavelengths, it can more easily navigate the nooks and crannies of urban environments.

UHF is commonly utilized for digital modes, satellite communication, and expanding networks by connecting repeaters. It provides clear, dependable voice, video, and data transfer connections, making it especially well-suited to contemporary digital communication techniques. UHF provides many opportunities for dependable, local communication and experimentation for amateur radio operators in highly populated locations.

Knowing the differences between HF, VHF, and UHF is similar to knowing the many climates and terrains you might encounter on a globe tour. Every band has its own distinct qualities, difficulties, and possibilities. Depending on your hobbies, location, and preferred mode of communication, one frequency appeals to you more than the others as your Ham Radio experience develops.

Recall that the bands serve as entry points to various Ham Radio communities rather than only being technical specs. You can communicate with far-off people and cultures on HF, with your local Ham community on VHF, and with state-of-the-art technology and urban exploration on UHF.

The journey includes investigating various frequency ranges, which present countless opportunities for connection, learning, and exploration. You're not just scanning frequencies when you adjust your radio throughout HF, VHF, and UHF—you're traversing a wide and diverse communication landscape. Greetings from the fascinating diversity of Ham Radio, where each frequency band invites you to embark on a new adventure of discovery and communication.

Modes of Communication: Voice, Morse Code, and Digital

In the vast realm of Ham Radio, you'll find a diverse range of communication options, each with its own appeal and difficulties. There's a place for you whether you find yourself drawn to the human touch of speech, the rhythmic dance of Morse code, or the cutting-edge world of digital modes. Let's explore these modes and see how they can improve your experience and help you connect with people worldwide.

Imagine turning on your radio, selecting a station, and hearing a person's voice across the street or in a different nation. Voice communication is the most straightforward and intimate method of communication. It is comparable to an international phone call in which the range of your signal is the sole restriction. The main voice modes used in Ham radio are SSB (Single Side Band) and FM (Frequency Modulation).

FM is the preferred method for local and regional communications, particularly on the VHF and UHF bands. Because of its clarity and ease of use, it's ideal for regular conversations, joining neighborhood networks, and communicating in an emergency. The preferred long-distance communication method is SSB, primarily utilized on the HF bands. While tuning and listening with more dexterity is needed, there's no comparison to the excitement of making a "DX contact"—a distant communication.

Voice communication offers real human connection, not merely the exchange of signal reports and position data. With the power of your voice, you'll establish friendships, tell tales, and give guidance.

Digital modes originated with Morse code, called Continuous Wave (CW). In Ham Radio, it may appear to be a thing of the past, yet it is alive and well. Operators use dashes (long beeps) and dots (short beeps) to communicate over the airways. This allows them to break through noise where spoken messages cannot and frequently use less power.

Acquiring proficiency in Morse code resembles picking up a new language, providing access to a world of efficient and minimalistic design. It takes work, but the payoff is a sense of achievement and a key to the polite and encouraging Ham Radio community. CW operators frequently comment that Morse code has a rhythm, almost like music, that makes it very fulfilling.

Ham radio is an evolving technology. Digital modes encode, transmit, and decode messages by utilizing the power of computers and software. Due to their extreme efficiency, these modes frequently allow communication even when signals are too weak for human perception.

Popular digital modes include PSK31, which facilitates longer keyboard-to-keyboard conversations; JT65, optimized for low signal situations; and FT8, which enables rapid, automated exchanges. Although every mode has its own peculiarities and community, they can all handle difficult situations and span large distances using less power.

Digital modes offer a playground for exploration, not merely for the excitement of making touch. You'll delve into the software world, adjusting parameters and perhaps even helping create new modes. A combination of creativity and communication keeps the activity interesting and novel.

You'll experiment with all these modes when you first start using ham radio. Digital modes are the frontier of technical inquiry, voice provides warmth and immediacy, and Morse code is a timeless relationship rooted in tradition. Depending on your interests, the environment, and the audience you're attempting to reach, each style has a proper time and place.

The appeal of Ham Radio is that it's a hobby that suits all types of people. Various modes are available, each providing a unique means of interaction, education, and discovery. As your skills develop, you may lean more toward one mode than the others, or you may just keep enjoying the diversity and switch between voice, Morse code, and digital, depending on your mood.

Recall that the mode you select expresses more about your interests and personality than it does about being a means of communication. Whether you're using voice communication to share stories, learning Morse code, or interacting with the digital world, you're a member of a global community dedicated to radio communication.

CHAPTER 3

Getting Started With Ham Radio

Entering the world of Ham Radio is an exciting adventure that offers a wealth of knowledge and a doorway to a global community. This chapter is meant to walk you through the process of starting, step by step so that your journey is fulfilling and pleasurable. Let's embark on this journey and embrace the spirit of exploration and community that characterizes amateur radio.

Getting a license to use a ham radio is the first step. This is your pass to enter the realm of amateur radio; it's not just a formality. Through licensing, all operators are guaranteed to possess the fundamental knowledge required for responsible and safe use of the airwaves. But don't let that scare you away. The goal of the procedure is to prepare and educate you, not obstructing you.

As they can differ, start by learning about the licensing requirements in your nation. Generally speaking, there are several licensing levels, each giving you access to additional frequencies and power. Start with the entry-level license, which covers the fundamentals of operating procedures, legal considerations, and radio theory. Numerous towns provide study aids, online resources, and workshops to aid with preparation. Recall that all licensed ham operators began exactly where you are right now. You're going to be a part of a welcoming and encouraging community.

Now that you have your license, it's time to select your first radio. Even though there are many possibilities available, it all depends on what you want to look into first. With its portability and ease of use, a handheld transceiver (HT) is an excellent place to start for VHF/UHF communication. A desktop transceiver might be your objective if you're driven by the appeal of long-distance HF communication, even if it necessitates a larger investment in gear and antennas.

Think about beginning with something easy and reasonably priced. Choosing the most sophisticated model for your first radio is unnecessary. Learning how to use your radio, establish contacts, and explore the bands corresponding to your license level is essential. Upgrades are possible as your interests and abilities grow.

In and of itself, setting up your first station is an adventure. The secret is to start small and grow as you acquire experience, whether building a desktop setup or a handheld radio. Your antenna is just as crucial to HF operations as your radio, if not more so. Be bold, take on DIY projects, and research the antennas that suit your hobbies and space. A satisfying aspect of the pastime might be creating and fine-tuning your antenna.

You will spend many hours exploring, tuning into frequencies, and establishing contacts at your station, which will serve as your own command center. Make it cozy and uniquely yours. Recall that the amateur radio community enjoys exchanging knowledge. Be bold and seek assistance when configuring your station or troubleshooting problems.

Making their first contact, or QSO, is a significant turning point in any ham's journey. Although it may seem intimidating, the town is renowned for its friendliness and hospitality. When you're ready, hit that transmit button after listening to the bands and getting a sense of the protocol. For novice operators, start with a local net or a calling frequency. When you introduce yourself as a new Ham, people will welcome and support you.

Remember that every operator you speak with has been in your exact situation. The goal of making contacts is to generate a network of friends, mentors, and eventually mentees, not only to trade signal reports.

Ham radio is a passion for lifetime learning. There's always something new to try, a band to check out, or a project to work on. Explore the accessible materials, including books, clubs, events, and internet forums. Take part in competitions, go on field trips, or attend your neighborhood amateur radio club meetings. These actions will help you become more integrated into the Ham community while expanding your understanding.

What makes Ham Radio so significant is its companionship, guidance, and respect for one another. Participate in the community by telling people about your experiences and listening to theirs. You'll eventually discover you're an important part of a worldwide family rather than just an operator.

Beginning a Ham Radio adventure offers countless opportunities for discovery, education, and interpersonal interaction. With every frequency you tune into and every contact you make, you are adding your own thread to the beautiful fabric of amateur radio. Greetings from the community; an exciting journey is ahead.

The Role of Amateur Radio Operators

When you enter the ham radio world, it's crucial to recognize the diverse functions amateur radio operators play. You're not simply taking up a new pastime; you're becoming a part of a worldwide society that conducts scientific research, communicates during times of need, and promotes goodwill among nations. Let's examine these responsibilities and see how your new path fits into the bigger picture.

One of an amateur radio operator's most important responsibilities is being a lifeline in an emergency. Conventional communication networks, such as landlines, cell phones, and the internet, can become overloaded or completely collapse during natural catastrophes. When this happens, amateur radio operators step in to bridge the communication gap and become the unsung heroes.

When equipped with a radio, you can use it to comfort impacted individuals, coordinate rescue operations, and transmit vital information to emergency agencies. Ham Radio's ability to function without the grid guarantees that assistance can be requested even in extreme situations. By accepting this responsibility, you become an essential component of an international emergency response network rather than just a hobby.

Technical experimentation is a playground in amateur radio. Globally, radio enthusiasts constantly invent and explore, expanding the limits of radio communication. Amateur radio operators have made enormous contributions to the world of communication technology, from creating new digital communication modes to bouncing signals off the moon.

As your interest in your pastime develops, you'll have the chance to learn more about the nuances of radio technology and share your own inventions. The Ham Radio community values This spirit of exploration, which encourages and supports those who go into uncharted territory in science and technology. In addition to deepening your understanding, your experiments and discoveries push the discipline forward.

Another aspect of ham radio is educating the next generation of enthusiasts and dispersing knowledge. Many operators take tremendous satisfaction in sharing their knowledge, teaching newbies, and helping them navigate the challenges of radio communication. The mentorship tradition guarantees the hobby's continued growth and vitality.

Participating in teaching programs, whether at schools, clubs, or open demonstrations, aids in deciphering the science and technology underlying radio transmission. It ensures that the hobby survives for future generations by encouraging people of all ages to learn more about Ham Radio.

More intimately, ham radio provides a special chance for cross-cultural dialogue and promoting world peace. You can communicate with someone from a different culture or nation whenever you use the radio. These talks can heal divisions, increase understanding, and create enduring connections.

Political and physical barriers cannot separate the spirit of Ham Radio, which unites a global village of operators via a shared passion for the hobby. Participating in our community builds a culture of understanding and goodwill, one conversation at a time.

In addition, ham radio operators are essential for planning and participating in competitions and events. These events offer the chance to hone your station, test your abilities, and interact with the larger Ham community in addition to being competitive. These events provide excitement and a purpose to the hobby, ranging from field days that simulate emergency operations to DX contests that test your ability to make long-distance connections.

By participating in these events, you and others will push yourself to reach new heights while contributing to the collective knowledge and expertise of the Ham Radio community.

Remember that as you start your Ham Radio adventure, you take on a role beyond the individual. You play a crucial role in a community that makes a difference, whether experimenting with new technologies, mentoring newcomers, bridging cultural gaps, or volunteering during emergencies or international competitions.

Accept these roles with joy and pride. In addition to radios and signals, amateur radio is about the people you get to know, the information you impart, and the connections you make. Greetings from the amazing world of amateur radio operators, where the impact of your efforts can reach far beyond the airways.

Equipment Overview: Transceivers, Antennas, and Power Sources

Starting a Ham Radio hobby is an exciting journey full of exploration and education. One of the first steps in this journey is learning about the tools that will allow you to reach out and touch the world. Let's review the essential equipment, which includes power supplies, antennas, and transceivers. Understanding these resources can improve your communication ability and increase your enjoyment of the amateur radio industry's art and science.

Envision grasping your initial transceiver, the central component of your amateur radio station. With the help of this adaptable gadget, you may send and receive messages over the airwaves by combining a transmitter and receiver into one device. Transceivers are available in various configurations, from lightweight handheld units ideal for local VHF/UHF frequency chats to more robust desktop models for long-distance HF conversations. Every kind has a distinct function and complements certain facets of the pastime. There is a transceiver for everyone, regardless of whether you want to connect with people nearby or across countries.

Selecting your first transceiver is similar to discovering the key that opens a treasure trove of international discussions. It's critical to think about the areas of Ham Radio that most interest you. Are you excited about making far-off, exotic connections on the HF bands? Or isusing VHF and UHF for emergency communication and community participation better? Your decision will be based on your response, but remember that you can use your first radio for a while. The vast nature of Ham Radio makes it a beautiful hobby with endless opportunities for growth and exploration.

Now, let's focus on antennas, the unsung heroes of radio transmission. Your transceiver's companion is an antenna, which transforms electrical energy into radio waves and vice versa. Transforming a whisper into a yell over the airways is possible with the correct antenna, which can significantly improve your capacity to send and receive signals. The designs of antennas differ greatly; basic wire antennas for HF bands are strung between trees, whereas tiny Yagi antennas for VHF/UHF are installed on rooftops. Everybody has their own traits, advantages, and disadvantages.

Building or installing your first antenna is a rite of passage because it's a practical introduction to radio wave physics. It's a request to try new things and gain experience through doing. Tuning your antenna or improving its capacity to receive and transmit signals, whether you choose a commercial model or go DIY, is a key component of the ham radio experience. Many hams feel that their enthusiasm for the activity grows in this environment amidst the wires and wavelengths.

Let's finally discuss power sources. Although your station's transceiver and antenna serve as its brains, power is its vital component. Most desktop transceivers need a steady power source, which is often supplied by a separate device connected to the electrical in your house. Power supplies are used to convert the AC current from your wall outlet to DC, which your radio equipment needs. For portable operations, you can use solar panels or batteries to power your station from virtually anywhere, such as parks, isolated hilltops, or during power outages.

Knowing power means knowing how to operate your equipment efficiently, not just how to keep it running. It's understanding how to balance the strength you require and the power at your disposal to ensure your communications go to their intended location without a hitch. Your power source is the cornerstone of your operation, whether from the comforts of home or the open air. It makes it possible for every word, code, and digital packet to transit the airways.

As you embark on this adventure, remember that your tools are extensions of your desire and curiosity. The antenna is your ears and arms reaching out; the transceiver is your voice to the world; and the power source is the constant pulse that powers it all. They provide access to a world of inquiry, communication, and learning.

Greetings from the exciting world of amateur radio, where you can explore limitless horizons with your equipment as the first step. Accept the learning curve, have fun experimenting, and remember that every contact you make adds a thread to the enormous, international amateur radio tapestry—you're not just broadcasting signals.

Setting Up Your First Station

Establishing your first amateur radio station is similar to sowing seed in an endless field of opportunities. It's a process that combines technical proficiency with creativity and joy with learning. Remember this as you set out on your adventure: every seasoned Ham operator started exactly where you are now—at the start of a journey that will lead to many discoveries and relationships. Together, we will carefully navigate this procedure to ensure your first experience with amateur radio is entertaining and successful.

Think about the area where you'll place your station first. It suffices to be a space where you feel at ease and can concentrate; it need not be big or ornate. It might be a desk, a room section, or even a transportable arrangement you can use outside. Consider this area your command center, where you can explore the airwaves and connect with individuals worldwide. Give it your unique touch by customizing it to motivate you.

On this voyage, the next step is to select your equipment. A transceiver is the foundation of your station, so start there. Depending on your interests and license class, as a novice, you may choose a portable or a basic base station transceiver that focuses on either VHF/UHF or HF bands. Keep in mind that the ideal radio is the one that fulfills your present requirements and provides flexibility for future expansion of your interests.

An antenna is the voice and ears of your station. Its importance cannot be emphasized because it directly affects your capacity for signal transmission and reception. Versatility and simplicity are essential for your first antenna. A vertical antenna might be an excellent place to start for HF, a straightforward dipole, and VHF/UHF. Although installation may seem intimidating, it's a fun DIY project and a chance to learn about the physics of radio waves. You can learn about the effects of your local surroundings on signal propagation by experimenting with antenna placement and direction. This process can also be entertaining and educational.

One important but crucial component that needs consideration is powering your station. Make sure your power source is dependable and meets the needs of your equipment. For most novices, a steady and clean electricity-connected power source will do. To ensure a safe and effective operation, it's critical to comprehend your equipment's power requirements and plan appropriately.

Now that everything is set up, it's time to initiate contact. This is a turning point in every Ham's journey, full of excitement and anticipation. Take some time to listen before you send it. You can understand the band's etiquette and conversational pace by tuning into different frequencies. When you're prepared, inhale deeply, turn on the microphone, launch the digital mode application, and say hello. Recall that everyone in the community is kind, and every Ham you meet begins here.

Your closest friends during these early days will be curiosity and patience. There will be obstacles to overcome, such as figuring out band conditions and optimizing your gear. Every problem is a puzzle, and solving it advances your knowledge and abilities while providing fulfillment.

Participating in local and online Ham Radio communities can greatly impact your experience. Participating in forums or joining a club can yield friendships, support, and helpful advice. The passion for communication and adventure unites the diverse and large amateur radio community. You'll find friends and mentors in this community—people who are as curious as you are and want to see you succeed.

Remember that setting up your first station is simply the beginning as you continue to develop your pastime. Every interaction in amateur radio presents a learning opportunity, making it a lifetime adventure. Beyond the airwaves, the relationships, abilities, and friendships you create enhance your life in ways you may not have realized.

CHAPTER 4

The Ham Radio Community

You'll soon see that the people you meet along the way are just as important as the radios and technology when you dive further into the world of Ham Radio. Every member of the global Ham Radio community has a unique story, set of experiences, and passions, making up a dynamic and diversified tapestry. This chapter celebrates that community, offers advice on navigating its waters, and extends an invitation to partake in its rich service, friendship, and education traditions.

Picture switching on your radio and selecting a station where, through the static, voices from many countries come to life. These people have distinct perspectives and a common interest in this pastime, not just nameless signals bouncing about the ionosphere. You join this international family the instant you make your first contact because you share a bond that cuts over linguistic, cultural, and geographic divides.

The foundation of the Ham Radio community is mentoring and giving. It's typical for seasoned operators, or "Elmers," to mentor and assist novices. They are ready to return the compassion they showed by remembering their humble beginnings. You'll discover that the community is friendly and helpful, so feel free to ask for help or find an Elmer. Someone always wants to help or share their knowledge, whether you need assistance with antenna setup, understanding radio theory, or planning your next equipment buy.

Engaging in clubs and activities is another method of fully immersing the Ham Radio community. Local groups provide a sense of community, resources, and educational opportunities. They are the beating heart of the amateur radio community, planning training sessions, competitions, and field days. Getting involved in a group can help you meet people for life, broaden your horizons, and learn new things more quickly. For many hams, field days, in particular, are the season's highlight because they combine the excitement of erecting makeshift stations in the vast outdoors with the companionship of cooperating to achieve a common objective.

Some areas of the community where camaraderie is evident are contests and "DXpeditions," which are excursions undertaken to make radio contacts from isolated regions. They're about pushing yourself, improving your abilities, and adding to the community's excitement and energy—not simply competition or networking. Every interaction forged during these gatherings fortifies the ties within the Ham Radio family, fostering the creation of stories and memories that will be cherished for future generations.

The Ham Radio community has a strong dedication to serving others. Hams worldwide are prepared to step in and offer vital connections between affected communities and emergency services during emergencies, such as when natural disasters damage conventional communication networks. This volunteerism is a mark of respect for the neighborhood and a concrete demonstration of the importance of the pastime to society. Using your knowledge and resources to benefit others and participate in emergency communication operations is a potent method to change the world.

The community is active online through forums, social media, and specialized websites in addition to the media. These platforms provide more areas for networking, learning, and sharing. They enhance the on-air encounters by letting you stay in touch with the community, find out about forthcoming activities, and stay updated on the most recent advancements in the pastime.

When navigating the ham radio community, always remember that you are an active participant in a living, breathing ecosystem. No matter how little you can contribute, it makes the community richer. Talk about your experiences, help out those who are just getting started, and recognize one another's accomplishments as fellow hams.

Joining Ham Radio Clubs

Entering the realm of Ham Radio entails experiencing the excitement of discovery, the delight of education, and the coziness of a supportive community. Joining a Ham Radio group is among the best methods to fully immerse oneself in this activity. Clubs serve as the lifeblood of the amateur radio community, providing support, knowledge, and companionship. They are much more than just associations of individuals with similar interests. This is a trip to explore the friendships and experiences that make Ham Radio genuinely unique, in addition to the airwaves.

Envision attending your very first club meeting. The sound of friends laughing and telling stories greets you, along with a sense of excitement about the topic or endeavor for the evening. Here, you're not just a single person with a radio; you're a family member bound together by a love of communication and a desire to discover all the opportunities that Ham Radio offers.

Taking a step into a wider world is joining a Ham Radio club. Clubs frequently have regular meetings, which can be anything from casual get-togethers to formal events featuring speakers, lectures on many facets of the hobby, and practical workshops. These gatherings are excellent learning tools since they provide insights into subjects that could take months to fully understand. Club meetings offer a plethora of knowledge freely shared among members, whether they are interested in learning the ins and outs of building antennas, comprehending the complexities of radio propagation, or investigating new digital modes.

Clubs, however, provide more than just instruction. They serve as centers of action, planning the gatherings that give the pastime life. A few examples include field days, competitions, and special event stations. Participating in these club activities may turn your Ham Radio experience from an individual endeavor into a group journey. Field days, for instance, emphasize cooperation, problem-solving, taking advantage of the great outdoors, and maximizing the number of contacts made. With the help and encouragement of your clubmates, these are chances to put theory into practice, learn by doing, and experience the excitement of working under less-than-ideal conditions.

Clubs are essential for emergency communications as well. When traditional communication networks fail during catastrophes, many are connected to local emergency management agencies and offer vital communication services. Participating in these initiatives may be quite fulfilling since it allows you to put your abilities to work and positively impact your community. It serves as a potent reminder of the worth of ham radio and the significance of your learning abilities.

Clubs provide a sense of community beyond the formalities. They're locations where people make connections, find mentors, and receive encouragement for their queries. Clubs are where the spirit of charity and knowledge sharing that characterizes the Ham Radio community really comes to life. You'll discover a friendly atmosphere and a warm welcome regardless of your experience level.

So, how may one locate the ideal club for themselves? Investigate the clubs in your neighborhood first. Numerous clubs have websites or social media profiles where you can find out about their events, meeting times, and membership requirements. Never be afraid to ask questions; even better, go to a meeting or event as a guest. This is a fantastic opportunity to meet some club members and experience the place's vibe.

Recall that there is no one-size-fits-all approach to clubs. Everyone has a distinct personality and goal. Some might focus more on outreach and teaching, some on emergency readiness, and some on contesting. Spend some time selecting a club that shares your interests and aspirations for the pastime.

A big milestone in your amateur radio adventure is becoming a member of a Ham Radio group. It provides access to chances for friendship, education, and community service within the larger Ham Radio community. Clubs are the epitome of the hobby: a love of communication that unites people, a dedication to helping others, and the excitement of exploration. Greetings from the family: surrounded by like-minded individuals excited to explore the airways with you, your adventure into Ham Radio takes on new dimensions.

Making New Radio Friends

When you start a Ham Radio adventure, you're not only venturing into uncharted territory but a world full of opportunities to make new friends. Imagine that every turn of the dial or push of the key is a knock on a door, and behind it is a fellow enthusiast ready to connect, share, and learn. A key component of the Ham Radio experience is meeting new radio friends who add support, camaraderie, and delight to your adventure.

Your initial interaction may feel akin to taking a leap into the unknown, a mixture of nervousness and excitement. Remember that every voice you hear over the static has been there before, someone who knows what it's like to reach out and feel heard. The Ham Radio community is renowned for its kind disposition, always ready to extend a warm welcome and encouraging word to newcomers. Now, inhale deeply, turn on your microphone, and say hello. You're going to open the door to an international network of friendships.

As you establish these relationships, you'll discover that every discussion offers a chance to learn something new—not just about radio but also about other people's cultures, languages, and life experiences. Don't be afraid to go beyond the basic call-sign and signal-report conversation you might begin with. Tell them about yourself, your hobbies, and even your hometown. You'll be shocked at how frequently a quick exchange may turn into an interesting discussion that establishes the basis of a long-lasting friendship.

Meeting people who share your interests via nets or regular on-air gatherings is another fantastic option. Nets are available in various sizes and forms; some are devoted to technical subjects, some to informal conversations, and others to particular hobbies. Getting involved in a net gives you a sense of camaraderie with like-minded people in addition to connecting you with other hams. The call signs and voices become increasingly familiar with time, turning the airwaves into a friendly neighborhood.

Field days and contests provide yet another way to meet new radio buddies. These gatherings are about more than competition; they're about teamwork, mutual support, and joy among the Ham Radio community. Long-lasting relationships are formed when people work together (or wave by wave) to accomplish a common objective. You'll discover that the friendships made on these occasions are as fulfilling as winning a medal or certificate.

Keep in mind that friendship goes beyond the air in the world of Ham Radio. Many amateur radio enthusiasts take advantage of the chance to get together in person, whether it is at regional club meetings, international conferences, or even Hamfests. These in-person exchanges give the friendships made on the radio more depth, converting call signs into handshakes and QSOs into coffee-talk exchanges. Because your activity is worldwide, you have friends who live far away, which makes journeys an opportunity for reunions and new experiences.

It's also about giving back to the community that accepts you when you make new radio pals. As you develop an interest in the pastime, you'll come across chances to help others by imparting your wisdom, providing support, and maybe even turning into an "Elmer" for newbies. This cycle of friendship and mentorship sustained the vitality and expansion of the Ham Radio community.

Ultimately, the main goal of Ham Radio is to meet new radio buddies. It's the mutual thrill of a successful interaction, the comforting sound of a well-known voice breaking through the background, and the common enjoyment of an international pastime. Every new acquaintance contributes hue, sturdiness, and texture to the tapestry of your Ham Radio adventure.

Thus, keep an open mind and a curious attitude as you reach out across the airwaves and continue to investigate. The real gems of Ham Radio are the friendships you form along the road, which enhance your experience beyond the dial. Welcome to the community, where everyone you meet has the potential to become a friend and where the promise of connection fills the airways.

Participating in Contests and QSO Parties

Engaging in competitions and QSO gatherings is like entering a communication festival, as the radio waves hum enthusiastically, and the competitive spirit unites the amateur radio community. Picture yourself getting ready for this occasion, with your desk prepared, your abilities honed, and you're ready to take on the maelstrom of connections that lie ahead. This is an invitation to celebrate your passion with like-minded people worldwide, not merely a test of your skills.

There are numerous variations of QSO parties and contests, each with guidelines and goals. Some may set a goal for you to meet as many people as possible in a given amount of time, while others may concentrate on connecting with people in particular areas or through particular communication channels. Whatever the format, these gatherings are fundamentally about interaction: interacting with the community, the airways, and the excitement of Ham Radio.

The excitement mounts as you get ready to play. Preparing to enter the battle and tune into the contest frequency is exhilarating. It may seem overwhelming to get into a fast-moving stream during your first exchange in a contest or QSO party, but remember that every operator you'll communicate with starts somewhere. Your confidence increases with every interaction, and what initially appeared to be a confusing swarm of call signs and signal reports turns into a beat you can join in on.

The variety of attendees these events draw is one of their most lovely features. The radio is a mosaic of voices, each adding to the event's fabric, from seasoned contestants with well-tuned stations to newbies keen to make their first contest contact. You'll discover that respect and camaraderie equals a competitive spirit. Every QSO (contact) you make is, after all, a shared success story and a link in the enormous amateur radio network.

Participating in competitions and QSO parties is another great approach to hone your abilities. The game involves managing fast, effective exchanges, tuning into faint signals, and navigating through congested bands. You will learn by doing as the dynamic environment pushes you to be your best operator. With each event, you'll also become increasingly skilled and perceptive to the subtleties of communication that characterize Ham Radio.

However, it goes beyond the rivalry and the people you meet. These gatherings honor the pastime and provide chances to test your boundaries and discover novel facets of ham radio. You could try out a new mode, connect to a region you haven't connected to before, or surpass your own record for QSOs recorded. Participation provides a deep sense of success reflected in the journey you've traveled and the scores you obtain.

Not to mention the tales you will discover along the route. Every competition and QSO party broadens your experience bank with stories of late-night operations, surprising encounters, and the mutual thrill of the final countdown. These tales form a part of your Ham Radio story, which you may tell friends and fellow club members and maybe even retell in contests down the road to inspire people who are just getting started.

These competitions bring the Ham Radio community together in ways that go beyond the excitement of winning. They remind us that, despite differences in geography and culture, we are all riding the same waves and sharing the same passion. The contest and QSO parties create links far beyond logged contacts; they are bonds of understanding, mutual obstacles faced, and joyous celebration.

So, keep an open mind and a spirit of adventure as you note the date for the upcoming contest or QSO party on your calendar. Whether your goal is to climb the scoreboard or simply network, remember that your involvement makes the event more lively and exciting. Greetings from the thrilling realm of competitions and QSO gatherings, where each call is an invitation to interact, engage in competition, and commemorate the remarkable Ham Radio community.

CHAPTER 5

Licensing: Why and How

As you venture into amateur radio, you may be curious about the need for a license and the process for obtaining one. This chapter seeks to demystify the licensing process by explaining its significance and walking through the procedures involved in becoming a fully licensed amateur radio operator. Your license is your pass to a worldwide network of creativity, collaboration, and exploration—it's more than just a legal authorization.

Let's start by discussing the "why." In ham radio, licensing has various important functions. It guarantees that all operators are familiar with the fundamentals of radio theory, safe operating procedures, and applicable laws. This information is essential for preserving the integrity of the frequencies and guaranteeing responsible and secure usage of the airwaves. It's also about respect: respect for the hobby, for other users, and for the regulations that preserve this exclusive mode of communication accessible to everybody. By getting your license, you're proving that you're committed to keeping these standards, which helps the pastime continue and flourish.

Although obtaining a license may seem difficult, it is intended to be attainable by individuals of all ages and backgrounds. Education is the initial step on the path towards becoming a radio operator. You can prepare yourself with a wide range of materials, including study guides, online courses, and in-person instruction offered by amateur radio groups. These materials cover everything from the basics of radio science and operation to the more intricate rules and etiquette. Learning can be easier and more enjoyable by approaching your studies with curiosity and an open mind.

Once you feel ready, the next step is to sit for the license exam. The exam's structure and content reflect each country's unique restrictions and band allocations. Multiple-choice questions are typically used in tests to gauge how well you understand your studied content. This is your chance to show off your newfound knowledge and convince the authorities that you're prepared to become an amateur radio operator.

It's a big accomplishment to pass the test. It's the result of all your effort and commitment and provides access to the wide world of amateur radio. Once you have your license, you are more than simply a person with a radio; you are a licensed operator with the legal right, expertise, and ability to use the airwaves for communication, exploration, and community service.

However, getting your license is just the start of your adventure. Amateur radio can lead to a lifetime of study and discovery. Your license lays the groundwork, but true learning happens from your experiences in the air. Every interaction is a chance to pick up new knowledge, hone existing abilities, and broaden your comprehension of the principles and practices of radio communication.

In addition, the licensing procedure gives you access to the larger amateur radio community. In study groups and exam sessions, you'll meet other enthusiasts who share your interests and passion. These initial contacts may result in friendships, mentorships, and a feeling of acceptance into a worldwide family bound by a shared passion for amateur radio.

Lastly, remember that having a license is a privilege with obligations. It is your responsibility to operate with civility and concern for others, follow the rules and band plans, and use the airwaves ethically. A licensed amateur radio operator participates in a tradition of service, communication, and research beyond sending out signals.

Getting your license to operate amateur radio is a worthwhile endeavor. It connects you to a global network of like-minded people and gives you access to a world of opportunities through a pastime. Suppose you approach the process with passion and commitment. In that case, you'll realize that the license you obtain is a key to a lifetime of learning and relationships rather than just a piece of paper. Greetings from the amazing world of amateur radio, where getting a license is just the start.

Understanding the Licensing Requirements

Knowing the license criteria is your first true step into this exciting world of Ham Radio as you stand at the beginning of your trip. It's normal to experience a mixture of anxiety and excitement, so let's dissect these prerequisites and transform the unfamiliar into the known. This is about more than just complying with regulations; it's about starting a global community and pursuing a journey of discovery.

In amateur radio, obtaining a license is a gateway, not just a formality. It guarantees that everyone holding a call sign knows the fundamentals of the science and manners that maintain the airways accessible and pleasurable for anyone. Although each nation has its own regulating organization and set of rules, the fundamentals of amateur radio licenses are always based on respect, safety, and education.

Getting to know the licensing body in your nation is the first step towards comprehending licensing criteria. This could be Industry Canada in Canada, the Federal Communications Commission (FCC) in the United States, or comparable agencies abroad. These organizations oversee frequency distribution, establish operation guidelines inside respective purviews, and specify the licensing procedure. They are the main source of current, accurate information about obtaining an amateur radio operator license.

Most nations organize their licensing procedures into a series of classes or levels, each conferring progressively more rights regarding output power and frequency access. An entry-level license is the first step in the process, introducing you to the fundamentals of amateur radio operations, including its legal framework, operational procedures, and technical features. The purpose of this foundational license is to enable you to enter the amateur radio community safely and efficiently.

You have the chance to learn and develop while you get ready for your license exam. You're learning things to improve your enjoyment of the pastime, not just stuff to cram for an exam. Several resources are available, including official study guides, online courses, and classes provided by nearby amateur radio groups. These resources cover the required information range, from an understanding of the electromagnetic spectrum and safe operation principles to radio theory and component function.

Exam day may feel intimidating, but remember that it's a step, not a barrier, in your journey. Be curious in your study and confident in your ability to pass the test because each right answer will familiarize you with amateur radio. As soon as you pass, you become a part of a group of people who have traveled the same route and are eager to learn, interact, and give back.

Once your exam is passed, you will receive a call sign, your exclusive designation in the amateur radio community. This call sign is more than a string of characters; it's your digital signature, a representation of your accomplishment, and your pass to global communication.

It's critical to remember that a license entails responsibilities. As a steward of the airwaves, you must conduct yourself with honor, decency, and care for other hams. The frequencies you use and the signals you send are part of a global communication network shared by all operators and depend on their cooperation and courtesy.

Therefore, obtaining a license requires more than merely navigating regulatory hoops. Licenses serve as the cornerstone of an activity that emphasizes connection and community just as much as technology and transmission. By accepting and comprehending these conditions, you're not just getting a license; you're also getting access to a community that will always welcome you, a world of limitless possibilities, and a pastime that will excite and challenge you for years to come.

So, confidently take the first step. Explore the resources, interact with the community, and regard the license application as the start of a once-in-a-lifetime experience. The world of amateur radio is waiting, brimming with new experiences, friendships, and the sheer delight of initiating communication. Greetings from a lifelong learner, where each frequency listened and call sign recorded represent a new discovery in the amazing world of amateur radio.

Types of Ham Radio Licenses

As you set out on your Ham Radio adventure, you'll soon come across a license roadmap, each offering a set of benefits and exploring opportunities. Consider these license keys to various doors inside the enormous amateur radio palace. Some doors lead to huge balconies with views of international communications, while others enter rooms teeming with local conversation. Understanding different license kinds is the first step in exploring the fascinating world of amateur radio. Together, let's navigate them with clarity and excitement for what's on the other side of each door.

Your first key is the entry-level license, also known as the Technician class in some other countries. It is intended to acquaint you with the essentials of amateur radio, encompassing fundamental laws, operational procedures, and technical know-how. This license is perfect for those who wish to engage in local and regional VHF and UHF radio communication, as it provides access to all amateur radio frequencies above 30 MHz. With this license, you can participate in amateur radio satellite broadcasts, join local nets, and experiment with Morse code communications on select HF bands. Think of it as the lower level of a mansion, where you can connect with like-minded individuals, learn valuable tips, and confidently embark on your adventure.

As your passion for it develops, you may want to travel farther and visit new lands and seas. The General class license is useful in this situation. It unlocks new segments of the global HF bands, much like a master key to many more rooms. You can join in international contests, communicate over long distances, and enjoy the excitement of DXing—making contact with far-off stations—all made possible by this license. Gaining a deeper comprehension of electronics, radio theory, and regulations, strengthening your foundation, and developing your skills are all part of preparing for the General class license.

The Amateur Extra license is available to individuals enthralled with the boundless possibilities of amateur radio and who want the most comprehensive access to all amateur bands and modes of operation. It is the key to the tallest tower in the home, which offers unparalleled privileges and views. It takes a deep understanding of sophisticated radio theory, rules, and operating procedures to reach this level. Though it's difficult, completing it will grant you the most spectrum privileges, enabling you to use exclusive frequencies, take the lead in emergency communications, and perform community service.

Any license is more than simply a collection of rights; it's an achievement in your Ham Radio journey, a testament to your expanding knowledge, proficiency, and involvement in the community. The path from Technician to General to Amateur Extra is one of ongoing education and inquiry, with each successive license opening up new frequencies and avenues for investigation, interaction, and contribution.

Whatever level you select, getting your license involves research, testing, and practice. It's evidence of your passion for the hobby and your resolve to become an informed and responsible operator in the international Ham Radio community. Remember that a network of mentors and enthusiasts is ready to support you throughout the licensing process. You're not alone on your path; there are study aids, internet resources, local groups, and Elmer support available.

While deciding which license to pursue, consider what areas of amateur radio most interest you. Do you find local community engagement and emergency communication intriguing? The Technician license could be the ideal place for you to start. Do you want to use your voice or code to connect people anywhere? Afterward, those doors will open for you with the General license and, eventually, the Amateur Extra license.

Every kind of Ham radio license is ultimately more than simply a certificate; it's a pass to a world of communication and a badge of honor that denotes your position in an international community of friendship, technology, and discovery. Whether you're just getting your Technician license or want to go to Amateur Extra, the licensing process offers a path of personal development and limitless opportunities. Greetings from Ham Radio's friendly and varied world, where obtaining a license opens up new opportunities for travel, friendship, and experience.

The Path to Getting Your License

Starting the process of getting your Ham radio license is a thrilling adventure that is full of anticipation, learning, and discovery. This route serves as your entry point into the hobby of amateur radio, where you can connect with like-minded individuals passionate about communication and discovery. Let's walk this route together, ensuring that each action you take is wise, self-assured, and ultimately fruitful.

Picture yourself starting this adventure to obtain your very own call sign, a badge of honor designating you as a part of the worldwide Ham Radio family. Although the process appears complicated initially, it consists of several doable steps that take you one step closer to realizing your dream.

Your first move should be to determine your country's license requirements. These specify the knowledge and abilities you have to show to obtain your license, and they are defined by the national regulating body in charge of amateur radio. It's critical to comprehend the several licensing classes that are accessible, as each has unique rights and obligations. You can adjust your preparation and make specific travel goals if you know what to expect.

Next, explore the plethora of study resources offered to prospective amateur radio operators. Resources are widely available and accommodate a range of learning styles, from official manuals and textbooks to online courses and instructional videos. Look for resources that will pique your interest in amateur radio while covering its technical and regulatory components. This phase aims to lay a strong knowledge base that will support all of your future amateur radio endeavors.

Interacting with the Ham Radio community is one of the most rewarding parts of your adventure. An abundance of seasoned operators referred to as "Elmers," are willing to guide novices by providing guidance, counsel, and support. Getting involved in internet forums or joining a local amateur radio club can help you meet others in the community, get answers to your queries, and build a sense of support and wisdom. This involvement is about joining a global network of friends and mentors, not just passing your exam.

As your knowledge expands, start preparing for the test. Many study books and online resources provide practice exams similar in structure and substance to the real licensure exam. These mock exams are really helpful in determining your level of preparedness, pinpointing areas that require more research, and boosting your self-assurance. See every practice exam as an opportunity to learn and a warm-up for the real thing.

It is now time to take the exam when you feel prepared. Nerves may accompany this phase, but remember that you've been ready for this. Knowing that you are prepared to demonstrate your knowledge and talents, go into the exam with confidence. The license exam is more than simply a test; it's a milestone that marks the completion of your perseverance and commitment to becoming an amateur radio operator.

Achieving exam success is a proud occasion and a significant turning point in your Ham Radio career. You will shortly receive your call sign, your special identification in the amateur radio community. This call sign is your ticket to a lifetime of discovery, communication, and excitement over the airwaves; it's more than simply a string of characters and digits.

Recall that getting your license is only the first step. Every interaction in the lifetime learning hobby of amateur radio presents an opportunity to learn something new, hone your abilities, and discover new aspects of the sport. Keep interacting with the community, participate in activities and competitions, and enjoy the countless opportunities amateur radio presents.

Obtaining a Ham radio license involves self-improvement, community building, and exploration. It's a route that opens up a universe of opportunities, encounters, and adventures in addition to a license.

CHAPTER 6

Preparing for the Exam

By preparing for your Ham radio license exam, you can join a global network of enthusiasts who share your enthusiasm for communication and adventure. It's an exciting journey to prepare for. Knowing that the road ahead is one of study, exploration, and the possibility of many global connections comforts you as you stand on the precipice of this new journey. Let's go over how to study so that you may go into the test with knowledge, confidence, and a genuine love for the pastime you will be testing.

Firstly, consider your preparation the start of your lifelong journey into Ham Radio, not just a way to get there. Your future experiences in the hobby will be built upon the subjects you study, which range from the fundamentals of radio theory to the complexities of operating procedures and laws. As an amateur radio operator, you get more tools with every idea and formula you comprehend.

Obtaining study materials is the first step in your trip. Many resources, including interactive study groups, video tutorials, and online courses, accommodate various learning styles. Look for study materials that pique your interest in Ham radio beyond just what is covered in the exam. Recall that gaining a solid knowledge base to enhance your experiences in the air is more important than simply passing the test.

Using the tools at your disposal, create a study schedule that works for your learning style and lifestyle. Assign yourself realistic goals and divide the curriculum into digestible chunks. Consistency is essential, whether you want to study for lengthy periods, a few days a week, or for shorter bursts each day. By providing structure and predictability to the preparation process, a study plan helps you stay on course and lessens anxiety.

Interacting with the community is one of the most satisfying components of studying for your Ham Radio certification. Many amateur radio clubs offer license classes and study groups that provide an excellent environment for learning. It's a great opportunity to ask questions, clarify doubts, and learn from others' experiences. In addition to the knowledge you acquire, this involvement is invaluable as it gives you a sense of community. It's not just about preparing for a test; you're also joining a worldwide community of radio enthusiasts.

Practice tests are incredibly important for exam preparation as they help you better understand the subject matter. They provide a realistic simulation of the exam experience by familiarizing you with the structure and types of questions you'll face. Additionally, practice tests help you to focus your study efforts more efficiently by identifying areas where you still need to learn. Treat each practice test as a learning opportunity, and continue to improve your understanding by reviewing your answers.

Although getting ready can be difficult, it's important to stay motivated and have an optimistic outlook. No matter how tiny your accomplishments have been, acknowledge them and remember why you started this road. Ensure you stay focused on your objectives, whether the excitement of establishing your first contact, the wish to support emergency communications or the delight of becoming a part of a worldwide community. The motivation that will propel you forward is your enthusiasm for Ham Radio.

On exam day, approach the test confidently. Have faith in the work you've done to prepare, the information you've acquired, and the network of people around you. Exam results are not just an evaluation of your memory; they also confirm that you are prepared to become an amateur radio operator.

Recall that finishing the test is only the first step. True learning happens with every dial turn, PTT button push, and new acquaintance you make on the radio. The license you obtain is your passport to a world of exploration, friendship, and development in amateur radio.

Getting ready for the test is just the beginning of an incredibly exciting journey in the amazing world of amateur radio. Thanks to your interest and dedication, you can connect people through the magic of radio transmission, and you can connect people through a passion that transcends boundaries.

Study Resources and Tips

Setting yourself up for success as you start your journey to become a licensed amateur radio operator includes having the appropriate study tools and following smart study strategies. Your chosen path is about building a solid basis for a pastime that can provide you with a lifetime of learning, exploration, and connection—it's not simply about passing an exam. Let's explore the world of study aids and advice that will help you prepare for the test and beyond while guaranteeing that every stage of the process is enjoyable.

Choosing study materials that align with your preferred learning method is the first step on your journey. A multitude of resources are available, each tailored to meet distinct needs and tastes. Start by consulting the official study guide issued by the amateur radio federation in your nation. These tutorials provide a thorough overview of the content you need to know and are specifically designed to meet the licensing exam requirements.

Online tutorials and courses can be a treasure trove for those who learn best by sight or sound. Many of these tools include interactive features, tests, and videos that simplify difficult ideas into digestible chunks. Because online learning is flexible, you may study at your own speed and review the material again as needed to ensure you grasp everything.

Local amateur radio clubs frequently provide study groups or license seminars for people who prefer teamwork. These environments offer structured instruction led by knowledgeable operators and give you an overview of the amateur radio world. The support and wisdom from other students and "Elmers" (mentor supervisors) can be quite inspiring.

Now that you have your study tools, the following step is to create productive study routines. Make a study plan that works with your everyday routine and set out designated periods for in-depth research. Maintaining consistency is essential; even brief, frequent study sessions can result in notable advancements.

Active learning strategies can improve knowledge and retention. Examples include making flashcards, teaching concepts to someone else, and summarizing information in your own words. Actively participating in the content helps you remember it better and can point out places that need more explanation.

Practice tests are a crucial component of your preparation strategy. They help you to identify your strengths and areas for improvement and familiarize you with the format and style of the questions you will encounter. Several online resources offer practice tests that closely resemble the exam environment and provide instant feedback on your performance.

Lastly, remember to look for yourself while studying. Ensure you're eating healthily, getting adequate sleep, and taking breaks to refuel. Maintaining general health is just as crucial as getting ready for the test.

As you proceed on this road, remember that every study hour you put in and every concept you grasp will help you get closer to being a member of the global amateur radio operator community. What makes this trip valuable is your desire to be a responsible operator, your curiosity about radio communication, and your passion for learning.

Understanding the Question Pool

Knowing the question pool is like finding a map that shows the landscape of your upcoming exam as you approach your objective of becoming a licensed amateur radio operator. This is a guide to help you navigate the terrain of knowledge you will encounter in your studies, not merely a list of possible exam questions. Let's examine this idea together so you can approach your licensure exam feeling assured and well-prepared.

All the questions on your exam are effectively stored in a database called the question pool. It is carefully created by professionals in the amateur radio community to cover all the topics you need to know about, including rules and regulations, technical issues, and operating standards. Consider it an all-inclusive synopsis of the material covered in the test, providing a clear picture of what you must understand.

Taking a closer look at the question pool can be very helpful. Not only does it allow you to test your knowledge, but each question also provides valuable information about amateur radio's theoretical and practical aspects. By studying the pool, you can understand how amateur radio works and why certain rules and practices are in place. This goes beyond just memorizing the answers.

Dividing the question pool into digestible chunks is a useful strategy. Instead of attempting to tackle them all simultaneously, take one topic at a time. By thoroughly approaching each subject, you can ensure you understand the fundamental ideas before proceeding to the next. It's similar to putting together a jigsaw in that each piece you place helps to reveal the larger picture.

An essential part of your preparation is practice tests, many of which include questions straight from the question pool. They let you assess your preparedness by acquainting you with the structure of the real examination and putting your knowledge to the test in circumstances close to what you'll encounter on exam day. Every practice test you finish is like a rehearsal, boosting your self-assurance and refining your precise and effective memory.

By participating in the question pool, you can also identify the areas that require more research. It is normal to run upon questions beyond your current understanding level or stump you. Instead of giving up, utilize these times as a guide to help you focus on the areas that need more attention. Your preparation will be productive and efficient thanks to this focused study approach.

Recall that becoming a skilled, informed amateur radio operator is the ultimate objective, not just passing the test. The question pool's vast variety of questions aids this larger goal. It motivates you to learn about every aspect of amateur radio, preparing you for the test and the real-world experiences you'll have on the air.

Let the question pool serve as a roadmap as you finish your exam preparation. You may confidently walk the path to licensure with the help of this resource, which provides clarity, direction, and understanding. Accept this as a learning and exploration experience, and remember that you'll get closer to being a member of the active amateur radio community as you answer more questions correctly. Welcome to a world of connection, invention, and communication, where the information you acquire now will set the stage for a myriad of adventures down the road.

Practice Exams and Effective Study Strategies

It's like navigating uncharted land when studying for your amateur radio license—incorporating practice exams into your regimen is essential. These tests are your compass, helping you navigate the wide terrain of knowledge needed for your journey; they are not just rehearsals. Together, we will explore the significance of mock tests and efficient study techniques to guarantee that your preparation is pleasurable and fruitful. Let's start this exciting journey.

Consider every practice test a mirror that reflects your areas of strength and regions where the terrain still needs to be clarified. With each question you answer, you're actively engaging with the content, stretching your comprehension and solidifying your knowledge, in addition to simply checking your recall. The allure of practice examinations is their capacity to replicate the real test setting, providing a preview of the difficulties and rewards you can expect on test day.

To get the most out of these useful resources, consider designing particular week periods for practice exams. As with the exam, treat them with the utmost seriousness and make sure you're in a peaceful, distraction-free setting. With this methodical strategy, every practice session becomes a potent learning opportunity that boosts your self-assurance and improves your test-taking techniques.

But passing a practice test is just the beginning of the adventure. Everyone presents a different chance for introspection and development. Examine your responses, paying close attention to the ones that gave you trouble. Examine in detail the justifications provided for both accurate and inaccurate answers. The goal is to find and fill comprehension gaps with credible, dependable information rather than ranking answers. Consider every error a learning opportunity rather than a setback on your way to a more thorough understanding of the subject.

Good study techniques go beyond mock tests and include an all-encompassing method of education. Divide the content into smaller, more digestible subjects to make a study plan that fits your lifestyle and learning rate. This deliberate, step-by-step method avoids overload and guarantees that every study session significantly contributes to your readiness.

Active learning strategies can help you get even more engaged with the content. Some examples of these strategies are making mind maps, flashcards or explaining concepts to a hypothetical audience. These techniques improve recall and add vibrancy and interaction to the learning process. They make the dull facts come to life by combining them into a story that stays with you long after you've put your books down.

Remember to include downtime and introspection in your study regimen. Taking regular breaks and engaging in mind-refreshing activities are essential. They guarantee that every study session is fruitful and guard against weariness. Remember that your brain requires time to assimilate and comprehend new knowledge; rest is essential on this journey rather than a luxury.

Using efficient study techniques and adding practice exams to your curriculum is essential for getting your amateur radio license. Learning about your preferred learning method and gaining self-assurance is as important as understanding the subject matter during this journey. With each practice test you finish and the topic you grasp, you're not just getting ready to pass an exam; you're also setting yourself up for a fulfilling journey into the exciting world of amateur radio.

Your greatest assets are your perseverance, curiosity, and dedication. Equipped with mock assessments and efficient study techniques, you're rapidly approaching the milestone of becoming a member of the global amateur radio community, prepared to investigate the frequencies and establish connections with like-minded individuals who value communication and discovery as much as you do

Exam Day: What to Expect

Exam Day is the day you've been looking forward to. This is the result of your commitment, diligence, and the several hours you've spent preparing; it's more than just an exam. It's normal to experience a mix of anxiety and excitement, so let's go over what to anticipate so you can face this day with assurance and composure.

Think of the morning of your exam. It's time to put what you've learned to the test and complete the necessary steps to become a member of the active amateur radio community. Getting a good start to the day is crucial. Eat a healthy meal and give yourself enough time to arrive at the exam site without hurrying. A serene start to the day establishes the mood.

Ensure you have everything you need before you depart, including your registration confirmation, a legitimate form of identification, and any resources or calculators that may be permitted per the exam guidelines. Packing water and some snacks for a short energy boost before the exam and a watch to help you keep track of time is a good idea.

When you get there, stop and take a deep breath. Recall that you are about to step into a world of opportunities in amateur radio; this is not just an exam. Volunteer examiners, who are also amateur radio enthusiasts and are aware of your adventure, will probably meet you. In addition to giving you the test, they also want to make you feel at home in the neighborhood. Please don't hesitate to ask them any last-minute queries you may have regarding the exam.

As soon as the test starts, thoroughly read each question. Thanks to your practice exams, you are ready for this. Have faith in your planning and expertise. Take your time with a question that stumps you. You can always come back to it later. Make a note of it and go on. Effective time management is essential to making sure you have enough time to provide your best response to each inquiry.

Keep your composure and remember to breathe during the exam. It's normal for anxiety to appear, but you'll be well-prepared and knowledgeable. You're getting closer to your objective with each question you answer. The exam validates your comprehension and readiness to enter the amateur radio community in addition to being a memory test.

After finishing the exam, consider the path that brought you here. Whichever way things turn out, you've acquired useful knowledge and abilities that will help you on your amateur radio travels. Examiners typically assess tests on the fly and give you feedback right away. A passing exam score marks a significant accomplishment in your amateur radio career.

Recall that there is still hope if the outcome is not what you had hoped for. It's only a diversion, a chance to review, absorb, and try again. Growth and tenacity are highly valued in the amateur radio community. All people begin somewhere, and obstacles are but a part of the path.

Exam day is an important turning point in your amateur radio career; it's a day when opportunity and preparation collide. Go into it with a clear head and faith in your skills. This test begins a lifetime journey into amateur radio, full of exploration, connection, and discovery. It's more than simply an exam. Greetings from the fascinating world of amateur radio, where today's difficulties will become tomorrow's successes.

🎁 HERE IS YOU FREE GIFT!

BONUS Q&A FOR HAM RADIO LICENSE EXAM

👉 [CLICK HERE TO DOWNLOAD IT](#)

A targeted collection of Q&As to bolster your preparation for the Technician Class Exam, enhancing your understanding and readiness.

CHAPTER 7

Operating Your Ham Radio

Taking a step into amateur radio with your recently obtained license is like unlocking an immense communication cosmos. The radio can transmit your voice, digital messages, or Morse code signals to other fans locally and internationally. To put theory into practice, this chapter will walk you through the thrilling process of using a Ham radio for the first time. Let's go on this trip together to make sure your radio debut is both fulfilling and pleasurable.

Assemble your station as a first step. It might be a more complicated configuration for international HF communications or a basic portable transceiver for local VHF/UHF band conversations. Recall that your station is your little island in the amateur radio universe, representing your goals and passions for the hobby. An antenna, a power supply, and a transceiver are the fundamentals. Ensure that the configuration complies with your license's requirements and any applicable local laws. You can grow and expand upon this basic setup as you become more involved in the pastime. It's only the beginning.

Every ham remembers their first contact, or "QSO," as a significant accomplishment. You should expect to feel nervous and excited simultaneously, but remember that every operator on the air has been in your position. Listen first. A feeling of the ongoing conversations can be obtained by tuning into a frequency. When you're ready to start speaking, wait for a lull in the conversation, turn on your microphone, and address the group using the manners you've studied. Your first communication, whether brief or drawn out, is an important milestone that marks your introduction into the worldwide amateur radio community.

There are several bands and modes available on amateur radio that you can experiment with. You can communicate using various options, ranging from instantaneous voice communications to the complex rhythms of Morse code and cutting-edge digital modes. The features of each band vary depending on the sun's activity, the weather, and the time of day. Try out several bands and modes to find out what excites you most. Your amateur radio license is a passport to limitless adventure, whether tracking down far-off contacts on the HF bands, participating in local nets on VHF/UHF, or decoding digital signals.

Being an active member of the amateur radio community is just as important as mastering the technology to operate a Ham radio. Become engaged in nets, join local clubs, and participate in competitions and field days. Engaging in these events is a great way to share your experiences, pick the brains of seasoned operators, and advance the hobby. The mentorship and friendship within the amateur radio community are well-known. You'll find camaraderie, support, and a common passion for scouting the airwaves by interacting with other Hams.

Consider each QSO as a teaching moment as you become accustomed to using your Ham radio. Amateur radio is a dynamic field with constantly emerging new modes, technologies, and practices. Remain inquisitive, and don't be afraid to try out novel configurations, antennas, or digital modes. The quest for knowledge and the openness to try new things makes the pastime interesting and new.

A responsible amateur radio operator is characterized by good operating etiquette. Always remember to listen more than you broadcast, be polite and patient, and respect frequency allotment when using the air. Keep in mind that amateur radio is a shared resource that connects people all over the world who share a passion for communication and exploration. By upholding proper etiquette, you support the integrity of the pastime and guarantee that the airways are accessible and fun for everybody.

When you first turn on your Ham radio, you're just starting on a journey that looks as diverse and rich as the airwaves. By pressing the PTT button and turning the dial, you are sending out signals and integrating yourself into a worldwide community. Greetings from the world of amateur radio, where there is always an opportunity for exploration, communication, and fun on the airways.

Making your First Contact

It's finally time to make your first radio contact using a ham radio. This important step will lead to a wealth of experiences and relationships in the amateur radio world. You're setting out on a journey that will open up a world of friendship, communication, and learning—it's not just about sending signals. Together, let's navigate this thrilling stage, ensuring you go into it feeling prepared, confident, and eager for the magic that awaits.

Picture yourself seated in front of your radio setup, with the gear you've laboriously selected and assembled ready to make it easier for you to join the worldwide Ham Radio network. There are countless opportunities on the airwaves before you, with every frequency as a possible link to a fellow enthusiast. It's normal to experience a slight trepidation combined with enthusiasm; all ham operators recall this time clearly since it was the lead-up to their first successful contact.

Give it some time to listen before you extend your hand. You may hear the distinct tones of Morse code, the buzz of conversations, and the digital sounds of different modes by tuning across the bands. The lively activity that exists among the amateur radio community is demonstrated by this symphony of transmissions. You may learn about the customs and cadence of band interactions by listening, which can assist you decide when and how to take the lead.

When you're prepared to commit, pick an obvious frequency or a sector known to welcome new members. Ensure your microphone settings are clear if you're speaking, and double-check your setup for accuracy if you're utilizing digital or Morse code modes. Next, inhale deeply, activate your microphone or begin your broadcast, and conventionally present yourself, usually mentioning your call sign and a succinct salutation. "This is [Your Call Sign], a new operator calling CQ and waiting for any calls," is an example of what it might say.

It's possible that your initial call won't get through right away, and that's acceptable, too. Operators from various time zones and locales tune in to amateur radio at their own pace. Here, being patience is not only a virtue but also a crucial step in the procedure. Try again if there's no response; if necessary, lower your frequency or select a different time of day when bands are known to be busier.

There is no greater feeling than hearing someone answer your call. Whether a fellow Ham lives halfway around the world or in the town next door, the airwaves suddenly reduce the space between you and them. Don't forget to have a pen and paper on hand to record crucial information such as their call sign, location, and any other information shared. This is the beginning of your logbook, a record of your travels through the amateur radio world, not just a formality.

Try to be as clear and succinct as possible throughout your initial conversation. It's acceptable to admit that you're new to the hobby; many operators in the amateur radio community are kind and willing to give novices guidance and support. This initial interaction is more than simply exchanging call signals; it's your introduction to a group of people who value dialogue, education, and helping one another.

Once you've finished your initial contact, consider the path that led you to this point. From preparing for your exam to assembling your workstation and placing this initial call, you've ventured into an entirely new world of discovery and interaction. From now on, every communication you have builds upon this first encounter; the airwaves are a never-ending source of learning, companionship, and adventure.

In Ham Radio, making your first contact is a milestone that unites you with every operator who has ever keyed a mic or touched a Morse code key. It's a combination of tradition and personal accomplishment. Greetings and welcome to the community, where making your initial move is just the start of a never-ending journey full of conversation, investigation, and the mutual delight of finding new things.

Operating Procedures and Etiquette

It's like studying the regulations of the road before you start driving when you start the thrilling journey of talking via Ham radio. You should also be aware of operating procedures and etiquette. These rules are not just about obeying the law but also about promoting a courteous, productive, and friendly culture among amateur radio operators worldwide. Now, let's explore the fundamental ideas that will direct your communications on the air, ensuring you develop into an important contributor to the amateur radio community rather than only an observer.

You should always begin by listening before you operate. Pay attention to the frequencies and take the time to understand how discussions proceed, the subtleties of exchanges, and how seasoned operators handle their correspondence. Listening is an active learning process rather than merely a passive one. It makes it easier for you to understand when to call, how to enter a discussion if necessary, and the general cadence of the band you're listening to.

Clarity and conciseness are your best friends when making calls, whether broad CQ calls (inviting any station to respond) or reaching out to a specific station. Pronounce your intentions and call sign clearly. For instance, the introduction "This is [Your Call Sign], calling CQ and listening" is a straightforward yet powerful method. Make sure the person on the other end of the phone has done speaking before you answer. They typically indicate this using the cue "over" or something similar. By doing this, the dialogue is kept flowing, and overlapping communications are avoided.

Technical accuracy and interpersonal interaction coexist in Ham Radio exchanges. Provide your name, location, call sign, and signal report (a gauge of how effectively your signal is being heard) at the outset. Every QSO (communication) starts with these details. Once you feel more at ease, you can personalize your conversations by disclosing information about yourself, such as your hobbies in amateur radio or other pertinent topics. But always consider other operators' time and the band circumstances; when the airwaves are crowded, keep conversations brief.

The amateur radio spectrum is a shared resource organized by band plans that specify which frequencies are assigned to voice, digital modes, Morse code (CW), and other forms of communication. It's important to follow these plans. By doing this, interference is reduced, and everyone can enjoy their time on the radio. Consider the band plan a commitment from the ham community to use the spectrum sensibly and politely.

Amateur radio will always involve some level of interference, whether intentional or brought on by band conditions. If interference arises during a QSO, respond to it politely. If another operator is involved, try to settle the matter politely. If unsuccessful, try modifying your antenna or frequency. Consider these opportunities to practice problem-solving and teamwork, as the amateur radio community cherishes both.

Every exchange needs signal reports because they offer insightful commentary on the transmission quality. The RST system (Readability, Strength, Tone) is the most often used format. By providing honest and accurate signal reports, other operators can increase the efficiency of their station by making necessary adjustments to their equipment or procedures. It's a mutually beneficial gesture that improves the experience of all band members.

Thank you should always come at the end of a conversation. The sense of community that distinguishes amateur radio is strengthened by acknowledging the other operator's time and company. Furthermore, regardless of your level of experience, always act gracefully in all your interactions. Mutual respect and support between participants are essential to the hobby's survival.

Adhering to these protocols and manners allows you to send signals and contributes to the culture of mutual aid, education, and respect that characterizes the amateur radio community. Your adventure over the airwaves is all about joining a global family brought together by a passion for communication and exploration, not merely finding new friends. Greetings from the amateur radio community, where each communication deepens the connections between hams worldwide and enhances your overall experience.

Using Repeaters for Extended Range

When you start out on the amateur radio journey, you'll quickly find that using repeaters opens up a whole new dimension of communication that enables you to reach farther into the distance and connect with other enthusiasts worldwide. Relays can be considered your pals in the air, amplifying your signal to increase coverage and improve your experience. Together, we will walk over the fundamentals of using repeaters so you can utilize these effective tools to the fullest with confidence and consideration for the community.

In essence, repeaters are radio relay stations erected strategically—often atop tall buildings or mountains—to receive signals from one amateur radio operator and retransmit them from a more advantageous position and at a higher power. This procedure greatly increases the communication range between operators, particularly when utilizing line-of-sight VHF and UHF frequencies. When you talk into your radio, picture your voice traveling over valleys and hills to reach receivers far away from your direct line of sight. That is what repeaters of magic provide to your adventures with Ham radio.

Finding the repeaters you can access is the first step in using them. Repeaters can be found listed by location, frequency, and access requirements in directories and online resources. Local amateur radio clubs can also be a great source of information; they can advise you on the most popular repeaters in your area and any special rules they may have. It's similar to sketching out the routes you'll go in the enormous realm of amateur radio by taking the time to investigate and comprehend the terrain of accessible repeaters.

To use a repeater, you must set your radio to the frequency of the repeater and any required access tones. A non-audible Continuous Tone-Coded Squelch System (CTCSS) tone is necessary to access many repeaters. The repeater receives this tone and your voice together, indicating that your call should be retransmitted. Consider it a covert knock that unlocks doors in the radio spectrum, enhancing and amplifying your signal's reach.

A code of behavior that guarantees these shared resources' peaceful and effective use is associated with using repeaters. Before transmitting, always listen to make sure the repeater isn't already in use. When you talk, ensure you know how long you are on the air by using your call sign as required by law. Community resources are repeaters, and sharing the airways with others is made possible by keeping conversations brief.

When you want to join a repeater chat, wait for a pause before introducing yourself. Being kind and thoughtful makes you seem good as an operator and creates a friendly and encouraging environment on the repeater network. Recall that collaboration and respect for one another are essential to the amateur radio community.

When using repeaters, it is important to consider technical factors. You need to be aware of the repeater's input and output frequencies and make sure your radio is configured to transmit on the input frequency and receive on the output frequency. To use a repeater successfully, you must understand the duplex operation, which involves using separate frequencies for transmission and receiving.

Additionally, be mindful of the clarity and power of your signal. Relays can be overloaded if you're too close and ineffectively accessed if you're too far away due to a weak signal. Achieving the ideal balance will improve your and your fellow operators' experiences by ensuring your messages are trustworthy and clear.

Using repeaters, you may connect with a larger community of amateur radio enthusiasts and hear your voice farther than you ever thought possible. Repeaters broaden your horizons by bringing far-off voices into your home, whether exchanging news, participating in networking, or just having a casual discussion.

Remember that repeaters are more than technical instruments when you use them in amateur radio operations; they are links between operators that allow for international friendships and relationships. Repeaters can be vital to your Ham Radio journey if you approach them with respect, knowledge, and a spirit of adventure. They open up new avenues for communication and investigation. Greetings from the extended amateur radio family, wherein repeaters help expand your community and make the world smaller.

CHAPTER 8

Exploring Advanced Topics

Your first license gives you many opportunities as you explore amateur radio. Exploring advanced issues, this chapter aims to broaden your horizons and stimulate your interest in the more complex and fascinating facets of the pastime. Together, with an open mind and a sense of adventure, let's set out on this voyage to explore the worlds that extend beyond the fundamental abilities of establishing contact and using your station.

Digital communication channels stand at the cutting edge of amateur radio, where tradition and technology collide. These modes have completely changed the way amateur radio operators communicate worldwide. They include FT8, JT65, PSK31 (Phase Shift Keying at 31 baud), and RTTY (Radio Teletype). Digital modes help you communicate in difficult situations with weak signal strength and interpret messages that would otherwise be lost in the noise. Entering the realm of digital communication broadens your experience in amateur radio while improving your technical proficiency and providing a more profound comprehension of how radio waves can be bent to achieve long-distance communication.

The excitement of establishing communication via an Earth-orbiting satellite is hard to imagine. For amateur radio operators, "satcom," or satellite communication, makes this fantasy a reality. Using a simple setup and some understanding of satellite orbits, you may communicate with hams on different continents and expand your reach beyond the horizon by organizing your interactions around the exact movements of satellites in the sky. Satcom forces you to think about radio communication's temporal and geographical components. Not only does this advanced topic increase your operational capabilities, but it also puts you in contact with the forefront of space technology and exploration.

DXing, or the quest for long-distance contacts and contesting, in which operators compete under strict guidelines to make as many contacts as possible within a predetermined amount of time, are considered by many to be the essence of amateur radio. With the thrill of tracking uncommon signals and the fulfillment of honing your operating techniques, these activities turn amateur radio into a worldwide scavenger hunt. By DXing and contesting, you get to know other enthusiasts who are as passionate about these exciting facets of the hobby as you are, and you also push yourself to learn more about propagation and enhance the performance of your station.

There is a special sense of accomplishment when good communication is achieved with the least amount of power. QRP operation, characterized by the use of transmitters with an output of 5 watts or less, indicates the equipment's efficiency and the operator's skill. This minimalist approach to amateur radio encourages a more creative approach to antenna design, operation, and overall station setup. It demonstrates that you can overcome obstacles and create valuable connections even with minimal resources if you have knowledge and creativity.

Through radio astronomy, amateur radio provides a means of accessing the cosmos and hearing the universe's whispers. This field includes using radio equipment to pick up signals from space, such as solar flares and cosmic background radiation. Amateur radio astronomy offers a practical method of investigating celestial occurrences, which blends scientific research with a pastime. It encourages you to use your amateur radio station as a means of exploration and wonder to further your understanding of the universe.

Amateur radio's use in emergency communication is among its most admirable uses. When conventional communication networks malfunction during a crisis, amateur radio operators offer vital connections between the impacted areas and emergency services. Attending emergency communication drills and training prepares you to assist your community in times of need and highlights how vital amateur radio is as a lifeline in an emergency. It serves as a reminder that the pastime is about more than just enjoying oneself; it's also about contributing to society.

Keep in mind that learning and exploring are ongoing aspects of the amateur radio hobby as you explore these complex subjects. Every new mode, satellite contact, and successful DX chase enhances your experience and relationship with the worldwide amateur radio community. Acquiring knowledge in these fields not only fulfills one's needs but also enhances the combined expertise and understanding of amateur radio operators across the globe.

You belong to a community that promotes innovation, discovery, and service, whether handling emergency communications, decoding digital signals, interacting via satellite, participating in competitions, or exploring the cosmos. We encourage you to push limits, discover new opportunities, and keep developing as an operator and contributor to this wonderful pastime in this chapter of your amateur radio experience.

Digital Modes and Software-Defined Radio

Within the fascinating field of amateur radio, the exploration of digital modes and software-defined radio (SDR) is a thrilling advancement that combines the rich history of ham radio with the most advanced current technology. This chapter aims to carefully lead you through these more complicated areas, revealing their enormous potential to improve your amateur radio experience while untangling their many intricacies. Accept this journey with an open mind and an open heart as we delve into the worlds of digital communication and the revolutionary potential of SDR.

With the ability to communicate in forms other than speech and Morse code, digital modes in amateur radio allow for new possibilities in text and image-based communication. These modes—FT8, JT65, PSK31, RTTY, and others—use computer software to encode and decode messages, enabling crystal-clear communication even in difficult situations where conventional means could break down.

Sitting at your station, picture yourself witnessing messages that travel great distances to reach you, from weak signals coming in from all over the world, turning into readable letters on your screen. Digital modes allow for precise and efficient image transmission, data interchange, and in-depth talks, so they're about more than just connecting the dots. They make it possible to investigate radio waves in ways that were previously unthinkable, converting noise into sophisticated communication.

Setting up your station with the required software is the first step towards preparing for digital modes. Numerous digital modes have programs specifically made with amateur radio operators in mind. These applications frequently have user-friendly interfaces that make it simpler for you to send and receive encrypted and decoded messages. Technical maneuvers may be necessary for setup, such as connecting your radio to a computer and customizing the software settings to fit your system. But as you go, you're improving your technical abilities and creating a link between your radio and the digital world.

One of the most lovely things about experimenting with digital modes is joining a thriving community of operators as passionate about this cutting-edge amateur radio as you are. Numerous learning, sharing, and networking platforms are available, including digital mode clubs, online forums, and special event stations devoted to digital communication. Other enthusiasts ready to share information, solve issues, and rejoice over successful digital QSOs (contacts) can be found here.

Software-defined radio (SDR) is a paradigm shift in the way we understand and use radio technology. SDR uses software to implement radio hardware operations like modulation/demodulation, mixing, and filtering. This adaptability allows a radio to become a powerful instrument for experimentation and exploration by enabling it to simply upgrade its software to support new frequencies, modes, and functionalities.

Using an SDR setup, you may tune between frequencies and observe signals dance across your screen to see the radio spectrum in real time. It's a fully immersive experience that gives you unmatched control and accuracy in your operations while improving your grasp of radio waves. SDR lets you control radio innovation, whether you're searching the airways for weak signals or experimenting with developing your own digital mode.

To get started with SDR, you only need to grab a USB dongle made for radio reception and some free software to get you started. Your SDR setup can develop into a sophisticated station with advanced receiving, analysis, and transmission capabilities as your interest and experience improve. With a wealth of courses, open-source software, and shared knowledge, the SDR community may assist you in becoming an expert by helping you navigate the transition from beginner to master.

Investigating SDR and digital modes does not imply abandoning the core principles of amateur radio. Rather, it's about extending the hobby's bounds and fusing innovation and history. With the help of these tools, you may interact, learn, and experiment in ways previously only possible in science fiction. These are instruments that improve your skills as an operator and help you better understand the countless opportunities inside the amateur radio spectrum.

As you go deeper into the worlds of software-defined radio and digital modes, always remember that every advancement brings us closer to the future of communication. You're a community member pushing the envelope of what's feasible by connecting people, exchanging knowledge, and delving into the mysteries of the radio waves that unite us all through technology. Welcome to this thrilling new phase of your amateur radio adventure, where tradition and innovation collide, and there's always something new to discover on the airwaves.

Satellite Communications and EME (Moonbounce)

As you continue to explore the intriguing realm of amateur radio, satellite communications and EME (Earth-Moon-Earth), commonly known as "moonbounce," are calling out to you as the next frontier of inquiry. With these complex themes, you may stretch beyond Earth's boundaries and touch the stars, providing a unique blend of technical difficulty and sheer awe. Embark on an extraordinary journey into the cosmos with the fervor of an explorer. Embark on a journey beyond the ordinary and explore the cosmos with passion.

With satellite communications, you can interact with people far away by creating a channel for your messages to travel across the sky. Imagine speaking up, having your code or digital signals bounce off a satellite, and then descending to meet another operator on the other side. This reality is available to anyone willing to learn how to use the amateur radio satellites that orbit our globe. Satellites have made it possible for this to happen; it is not science fiction.

Before you can begin, you will need to familiarize yourself with the many kinds of satellites available for amateur radio use. Digital satellites, FM repeater satellites, and linear transponder satellites are some of these satellites. Each one offers a unique means of communication; the simpler voice modes are achieved using directional antennas and handheld radios, while the more complex digital modes require specialized equipment. The rush you get from making your first contact with a satellite is unmatched; it's a breathtaking moment when aspiration and technology combine to conquer the vastness of space.

Comprehending satellite orbits and passing forecasts thoroughly is crucial. Many software programs and online tools can help you track satellites in real time and provide you with exact information about where and when to place your antenna. By using this knowledge, the seemingly insurmountable task of satellite communication becomes not only doable but also pleasurable.

Using the moon as a natural satellite, Earth-Moon Exchange (EME) is a communication mechanism that sends signals from one place on Earth to another. This technique elevates the idea of long-distance communication. EME is essentially about the joy that arises from successfully conquering one of the most challenging challenges in amateur radio despite the idea's initially frightening appearance.

Amateur Radio's EME communications sector blends advanced technical proficiency with a hint of celestial wonder. EME communications is the best option if you want to start out in this industry. The basic principle is sending a strong signal toward the moon, which will then reflect back to Earth and be picked up by another operator. Precise alignment, cautious timing, and a deep feeling of wonder are necessary for the dance between the Earth, the moon, and technology.

Although EME may require more sophisticated gear than other amateur radio operating modes, such as high-gain antennas, low-noise preamplifiers, and high-power amplifiers, there's no comparison for the pride one feels after successful moonbounce contact. This illustrates human ingenuity and the spirit of adventure that defines amateur radio operators as a group.

To delve into the world of satellite communications and EME, one must possess the requisite technical skills and an open mind that encourages experimentation and ongoing learning. Reaching out to the amateur radio operator community is a good place to start. Many operators are willing to share their knowledge as they specialize in these fields. Local clubs, special interest organizations, and internet forums can all offer invaluable advice when choosing the right gear and setting up your first satellite or EME encounter.

Part of your work should involve understanding the physics underlying satellite orbits and moonbounce events. Acquiring this foundational information will enhance your operational proficiency and broaden your awareness of the technical and natural wonders that enable you to interact with others.

Experiment with antennas and equipment on a modest scale at first, then increase it as your interest and skill level grows. In terms of satellite communications, a handheld radio with a directional antenna can also open up intriguing new avenues for connection. Moonbounce's marvels may be enthrallingly introduced to EME by starting with the reception of signals bounced off the moon.

Remember that as you study satellite communications and EME, you are getting involved in a kind of amateur radio that really pushes the envelope of what is feasible. All of the contacts you make and signals you bounce off of celestial bodies are a constant reminder of the vast potential within this activity. Not only are technological advancements significant, but human relationships forged over vast distances are also noteworthy, with the stars acting as a conduit for these exchanges.

DXing and Awards

Starting a DXing adventure on amateur radio is like starting out in a huge ocean of communication, where every new country or island is a treasure found, and every contact is a discovery. This chapter is devoted to helping you navigate the exciting world of long-distance communication or DXing and pursuing prizes that recognize and honor your accomplishments in using the airways to connect people, cultures, and continents.

The thrill of connecting with a station thousands of miles away and having your signal reach places you've only ever dreamed of traveling to is unimaginable. DXing is that link between the virtual and the actual world that lets you travel the world without leaving the comforts of home. It's not just about gathering contacts; it's also about the backstories behind them, the formed trans-frequency friendships, and the mutual thrill of cutting through the clutter and distance.

DXing forces you to sharpen your abilities, comprehend the subtleties of propagation, and constantly enhance the configuration of your station. Every choice, whether tuning your antenna to pick up the tiniest whisper from a far-off place or determining the ideal time of day to hear the bands come alive, is a step closer to creating that next remarkable connection.

Awards are significant events in the amateur radio community that recognize your commitment, expertise, and love of DXing. These accolades can range greatly, from being confirmed as having made contact with a certain number of nations to competing in particular competitions or establishing contact through various channels and bands. Each certificate or plaque is a concrete representation of your journey, telling a tale of endurance, patience, and the excitement of the chase.

The American Radio Relay League (ARRL) bestows one of the most sought-after accolades, the DX Century Club (DXCC), to operators who have successfully communicated with at least 100 nations. Many more awards are just waiting to be discovered and attained, each with its own special requirements and appeal. Your DXing experiences gain excitement and purpose as you pursue these rewards, encouraging you to stretch yourself and pay closer attention.

Many people liken DXing to hunting—a pursuit for that elusive or unusual contact that will add a rare jewel to your collection. The excitement is in the pursuit, in the suspenseful moments spent tuning into your radio and listening for a call sign that denotes a new nation or island. It's a hobby that pays off when you're patient and persistent. Occasionally, the stars align, and faraway signals come in strong and clear, allowing you to connect with another operator who may be halfway around the world.

Patience, strategy, and a thorough understanding of radio wave propagation are all necessary for successful DXing. Knowing when and where to listen is just as important as having a strong transmitter or an advanced antenna setup. Your signal's range is greatly influenced by the time of day, solar activity, and the seasons. As you become familiar with these patterns, you'll have the intuition to determine when it's ideal to start your DXing adventures.

The feeling of camaraderie that DXing fosters is truly special. Every interaction provides an opportunity to gain insights into a different culture, share stories, and make friends worldwide. This international sense of community is crucial to the amateur radio community since DXers regularly exchange advice, congratulate each other on their achievements, and inspire others to pursue those challenging contacts.

Remember that every contact you make when DXing and pursuing awards is an adventure and a step into the unknown. There will be days when the bands seem silent and days full of victories, but every day is an opportunity to learn and develop as an operator and a part of a worldwide family brought together via the airwaves.

The journey—the early mornings spent listening to the sunrise across the bands, the thrill of logging into a new country, and the delight of a hobby that brings the globe to your doorstep—is ultimately the real pleasure of DXing and award-chasing. Your enthusiasm for amateur radio, a pastime fundamentally about exploration, discovery, and interacting with the outside world, is demonstrated by your DXing and award-winning.

Greetings from the fascinating world of DXing and prizes, where every frequency is a doorway to new experiences, and every exchange honors amateur radio's ability to bring people together. In the hunt for far-off signals and highly sought-after prizes, you'll discover successes and obstacles and a pastime that enhances your life with each call.

CHAPTER 9

Ham Radio for Emergency Communication

You've entered the amateur radio path and found a world of intriguing conversations, digital adventures, and astronomical transmissions. However, Ham Radio's utility is among its most profound features in times of need. Turning the page to this chapter, "Ham Radio for Emergency Communication" takes you into a new world where your hobby becomes more than just a pastime; it becomes a lifeline for communities during emergencies.

Imagine a situation where landlines, cell phones, and the internet are no longer functional due to natural disasters like hurricanes, earthquakes, or severe storms. Traditional communication networks have collapsed. The amateur radio community shines at these times, facilitating vital contact between affected people, relief agencies, and emergency services. As an amateur radio operator, you play a vital part in helping others in need by serving as a beacon of hope and connectivity. Your radio is more than just a tool for discovery.

The secret is to prepare. More than just technical proficiency is needed for emergency communication; one must also possess a thorough comprehension of emergency procedures, the capacity to function under pressure, and the willingness to assist. Start by becoming acquainted with the frameworks created to facilitate emergency communication: the Incident Command System (ICS) and the National Traffic System (NTS). ARES (Amateur Radio Emergency Service) and RACES (Radio Amateur Civil Emergency Service) are local amateur radio clubs and emergency communication groups offering priceless training and expertise.

Your field toolkit is a portable emergency radio station or Go-Kit. Along with your radio equipment, it should contain any personal things needed for prolonged operations and necessities like extra batteries, power sources, antennas, and headphones. The goal is to be prepared to deploy quickly and effectively, whether from the field, an emergency operations center, or a shelter.

Taking part in exercises and simulated emergency tests (SETs) is essential. These drills, which are frequently led by neighborhood associations or emergency communication teams, simulate emergency procedures in the real world without the added stress of a catastrophe. They test your capacity to assemble swiftly, perform well under less-than-ideal circumstances, and collaborate well with others in an emergency response team. Every drill helps you hone your abilities, making you a more capable and self-assured operator in the future.

Adaptability is key when communicating in an emergency. Things can change quickly during an emergency, so you have to be quick-thinking and creative with your tools and methods. This could entail creating a makeshift repeater to increase the range of communication, making an antenna out of readily available materials, or even coming up with novel ways to transmit important data. Your inventiveness and adaptability can greatly impact how well the response effort goes.

The attitude of cooperation and community lies at the core of emergency communication. Collaborating with fellow hobbyists, you become a member of a strong network devoted to promoting the common good. Because of the common experience of making a good difference during difficult times, the relationships forged during these crucial periods frequently endure a lifetime. Recall that in emergency communication, your voice represents not just your own but also the voice of your community when it comes to asking for aid, expressing support, or offering consolation.

Managing during an emergency requires a high level of accountability. You must act with professionalism, decency, and a clear understanding of how serious the situation is. You should always follow the guidelines of truth, clarity, and patience since misinformation or misinterpretation can have detrimental effects. Furthermore, maintaining secrecy and privacy is crucial since you can be handling sensitive information.

Ham Radio can be incredibly useful for emergency communication, providing a sense of satisfaction for those who use it to positively impact someone's life. From enabling vital supply deliveries to reuniting families through welfare and health messages to issuing life-saving early warnings, amateur radio has proven that it is not just for recreational use. Its numerous benefits extend far beyond that, making it an invaluable tool for emergency communication.

When you use ham radio for emergency communication, you're not just a hobbyist but an essential component of the world's safety net, prepared to step in when the unimaginable occurs. Welcome to a chapter where your love of amateur radio combines with the highest call to duty, where every frequency tuned and a message sent highlights the vital role that amateur radio plays in protecting communities. This epitomizes what makes ham radio so powerful: the waves become the threads that bind a fabric of resiliency, encouragement, and hope.

Role of Ham Radio in Disasters

Amateur radio is a ray of optimism and resiliency in the face of calamity when the frail threads of traditional communication break under the weight of turmoil and uncertainty. Entering the world of Ham Radio has equipped you with more than simply a rewarding pastime; it has made you an essential part of an international emergency response network. Serving communities in their hour of greatest need is where the genuine spirit of amateur radio truly shines.

Consider the aftermath of a natural disaster: phone towers are unusable, electricity lines are down, and the internet is a thing of the past. The world seems broken and disjointed at these times by the might of nature. However, in this gloom, your abilities as a ham radio operator light a fire, creating bonds that break through the alone. Your station becomes a lifeline connecting emergency responders to the communities they work to protect and to each other beyond simple equipment.

Effective disaster response requires proper preparation. As a ham radio operator, you possess the skills to communicate clearly and quickly and knowledge of emergency procedures and equipment specifications. This planning is crucial as it enables you to act promptly when time is of the essence. You can also help coordinate rescue and relief activities by providing vital communications.

Catastrophes are unforeseen events that abruptly change the demands and challenges landscape. Your flexibility—to change frequencies, adjust your setup, or come up with last-minute fixes—becomes quite valuable. Here is where ham radio's adaptability shows off; it can run off the grid, run on batteries or generators, and get past obstacles like damaged infrastructure. This adaptability guarantees that communication is feasible even under the most difficult circumstances.

You have a more important duty to play after a crisis than just using a radio technically. Perhaps most crucially, you symbolize the community's resiliency and a calm voice amid a storm. You also become a source of information for those isolated from the outside world. You can reunite families that have been estranged by facilitating health and welfare traffic and informing loved ones about their safety and well-being. Every word you send uplifts the spirits of people impacted by it by bearing the weight of hope.

Working in concert with local, regional, and federal emergency services, ham radio operators frequently fill up communication gaps and enhance official channels. This cooperation is based on respect for one another and a common objective of protecting people and property. You can be responsible for transmitting vital information from far-off places, giving current condition updates, or assisting with reaction effort logistics coordination. By doing this, you become an essential member of the emergency management team rather than merely a volunteer.

The success of ham radio in emergencies results from ongoing group learning and personal preparedness. Engaging in exercises, role-plays, and educational programs alongside emergency management organizations and other amateur radio operators refines your abilities and expands your comprehension of the complexities associated with emergency communication. These experiences help you prepare for the psychological and emotional strain of working under pressure, in addition to the technical difficulties.

Every transmission in the center of emergency operations is guided by your ethical compass. When it comes to ensuring that the information transmitted is trustworthy and that operations support rather than impede response efforts, discretion, accuracy, and adherence to protocol are crucial. You have a great deal of trust as an operator, and that trust entails a duty to behave honorably and purposefully.

Reflection becomes a tool for growth after the storm has passed and the immediate crisis has subsided. Every operation and tragedy you've survived teaches you priceless lessons that improve your skills and the preparedness of the amateur radio community overall. By exchanging experiences, difficulties encountered, and solutions discovered, amateur radio operators can enhance their collective expertise and become more equipped for the next call to duty.

You represent the spirit of community, ingenuity, and service as you play the part of Ham Radio during calamities. Your donations cut across national borders, providing hope and a sense of community when needed. This leg of your amateur radio adventure demonstrates the unwavering spirit of people who are prepared to help in times of need and is a monument to the positive effects of communication. Greetings from an aspect of amateur radio that goes beyond the airwaves to change the world, one message at a time.

Getting Involved with ARES and RACES

When you first started using amateur radio, you learned about its ability to connect, explore, and—more than ever—serve. When you dig further into the core of this hobby, you discover that it provides a special way to give back through participation in the Radio Amateur Civil Emergency Service (RACES) and the Amateur Radio Emergency Service (ARES). These groups embody the dedication to community and preparedness that characterizes amateur radio. Let's talk about how you can use your love of amateur radio to make a significant difference in disaster relief and emergency communications.

ARES and RACES are two aspects of amateur radio that are extremely important in times of emergency. They use the expertise and commitment of amateur radio operators to guarantee that communications are maintained when they are most required. ARES, supported by the American Radio Relay League (ARRL), offers emergency communications to a wide range of agencies and nonprofits during disasters. RACES functions exclusively during civil emergencies declared by local, state, or federal authorities. It was created by the Federal Emergency Management Agency (FEMA) and is overseen by local, county, and state emergency management agencies.

Participating in ARES or RACES is about more than just using your amateur radio expertise; it's about helping your community in times of need. It's an opportunity to be a rock of support, ensuring affected parties and emergency responders have a trustworthy way to communicate when it matters most. In addition to being gratified to assist, it's an opportunity to collaborate directly with expert emergency management teams and acquire knowledge and expertise that can only be obtained via practical experience.

Serving others and being open to learning are the first steps in being involved. Operators for ARES and RACES must be skilled in amateur radio and comprehend the unique rules, guidelines, and standards of emergency communications. First, contact the ARES or RACES group in your area. These groups frequently provide training and orientation sessions for new members and always need committed volunteers.

Appropriate training is necessary to serve in emergency communications effectively. From specialist courses on emergency protocols and digital communications to basic amateur radio operations, ARES and RACES provide a range of training possibilities. Volunteers in emergency communications can take courses in incident management, emergency operations, and other topics through FEMA's Independent Study Program. Obtaining certifications in programs like the National Incident Management System (NIMS) and Incident Command System (ICS) can increase your efficacy and match your abilities to those of emergency responders in the field.

Participating in drills and simulations is one of the finest methods for preparing for real emergency situations. These drills simulate actual crisis situations and offer practical experience in equipment setup, high-stress operations, and teamwork in emergency response situations. Drills are a great way to evaluate communication strategies, pinpoint areas for development, and boost self-assurance in one's ability to function under duress.

Having a Go-Kit—a small, rapidly deployable assortment of personal items and amateur radio equipment—is crucial to being prepared to serve with ARES or RACES. Along with your radio equipment, your Go-Kit should have extra batteries, antennas, headphones, personal identification, and necessities for yourself. When disaster strikes, having a fully-stocked Go-Kit guarantees that you can act quickly and continue to function for long periods.

Enrolling in ARES or RACES requires a commitment: you promise to always be prepared, always learn and grow, and serve your community with professionalism and dedication. Technical proficiency, adaptability, and the capacity to collaborate with other volunteers and trained emergency responders are all necessary.

The real benefit of joining ARES or RACES is knowing that your efforts directly affect the safety and well-being of your community. Your assistance is greatly appreciated, whether it's communication facilitation during a natural disaster, search and rescue assistance, or life support during a public event. The intangible rewards that enhance your amateur radio adventure are the appreciation from those you've helped and the feeling of pride you have in making a difference in something bigger than yourself.

When you consider volunteering for emergency communications, remember that you are joining a distinguished heritage of amateur radio operators who have long served as first responders to disasters. Participating in ARES or RACES is more than just a way to further your hobby; it's a powerful way to demonstrate the three main principles of amateur radio: preparation, community service, and service. Embark on a fulfilling journey where your enthusiasm for amateur radio becomes a ray of hope and fortitude for people in need.

Setting Up an Emergency Communication Kit

The ability to have a significant impact during times of crisis is inherent in every amateur radio operator. Preparing an Emergency Communication Kit, often known as a "Go-Kit," is essential in this crucial hobby phase. If something unexpected takes place, this kit is not merely a collection of equipment; rather, it is your portable command center that is poised and ready to support essential communication at any moment. Let us guide you through putting together this indispensable toolkit so that you are well-equipped, well-prepared, and ready to help your community when it requires your assistance the most.

In times of crisis, your Go-Kit is more than just a piece of equipment; it is a ray of light and a connection to survival. Your gear helps you function efficiently, offering essential assistance to emergency services and affected individuals, regardless of whether the crisis is a natural catastrophe, a community event, or any other circumstance in which traditional communication channels fail. Being prepared is the essence of the Go-Kit; it encapsulates your capacity to react quickly and keep communications going even when faced with difficult circumstances.

Make sure you have the appropriate container. Make sure that your Go-Kit is portable, long-lasting, and resistant to the elements. It would be good to have a durable tote, rolling bag, or backpack to safely transport your stuff and survive difficult situations. Consider the item's size and weight; even when completely loaded, it should be manageable. In the event of an emergency, it is important to remember that you may need to move rapidly or travel through places littered with debris.

Your radio equipment is the most important part of your Go-Kit. Selecting a dependable, versatile radio that covers the bands and modes you will most likely activate in times of need is important. A mobile unit with a wider range can give increased capabilities, while a dual-band VHF/UHF handheld radio is the very minimum essential for local communications. It's important to ensure that your radio remains functional for as long as possible, especially in emergencies. To achieve this, it's recommended that you include a battery pack, extra batteries, or a portable power source like a solar charger. Additionally, clear communication can be achieved by using a high-quality headset or hands-free device, allowing you to have your hands free to perform other tasks.

To have successful communication, antennas are absolutely necessary. Suppose you want to extend the range of your radio. In that case, bring a portable antenna that can be simply deployed. A few different options are available, such as a roll-up J-pole for VHF/UHF operations or a small wire antenna for HF operations. Coaxial cables, connectors, and a robust rope for hoisting antennas into position are other vital components of the antenna installation system. To achieve the highest possible signal strength, you should incorporate an SWR (Standing Wave Ratio) meter into your antenna tuning setup.

The availability of dependable power sources is non-negotiable. Additionally, you should incorporate hand-crank generators, portable solar panels, or extra battery packs into your radio and the batteries running it. Power adapters and a multi-port USB charger can keep extra devices powered. Remember that the objective is sustainability; at the very least, you want to keep activities going for an extended period.

A basic tool set that includes screwdrivers, pliers, wire cutters, and electrical tape can be used to make quick fixes and modifications. To make more complicated repairs, you need a soldering iron and a limited selection of connectors and spare parts. When it comes to preserving communication capabilities during ongoing events, having the ability to troubleshoot and repair your equipment can make all the difference.

It is also important that your Go-Kit caters to your individual requirements. Be sure to bring clothes suitable for the weather, water, snacks that do not require refrigeration, a first aid kit, and any prescriptions they may require. In addition, it is essential to have a flashlight, additional batteries, a multi-tool, a notepad, and a pen to record contacts and make notes. Comfort and self-sufficiency make it possible to focus on the task at hand without being distracted by concerns about fundamental requirements.

It is important to carry copies of your identity, your amateur radio license, and any certifications pertinent to your profession, such as first aid or cardiopulmonary resuscitation. A basic operating guide or manual for your radio should be included, as well as a list of frequencies and repeaters specific to your region and information on who to contact in the event of an emergency. This will prove quite useful in high-pressure situations where memory may be impaired.

Putting together your Go-Kit is just the first step in the process. Become familiar with everything, and until it becomes second nature, practice setting it up and operating it. Putting your equipment through its paces in various environments can be accomplished by participating in field exercises, drills, and "radio in the park" days. The objective is to make certain that you are prepared to implement your station and communicate effectively in the event of an emergency and confident in your capacity to do so.

In conclusion, you should consider your Go-Kit a living thing. As your involvement in emergency communications increases, technology evolves, and you gain more expertise, your gear ought to evolve to accommodate these changes. You should perform routine inspections and updates on your equipment and continually look for methods to improve its usefulness and dependability.

By putting together your Emergency Communication Kit, you are not only becoming ready for the worst possible scenario but also embracing an essential component of the amateur radio philosophy, which is to assist other people. In times when it means the most, this kit represents your preparedness to stand up and deliver a crucial service to those who require it. Your talents, equipment, and passion join together to make a difference in amateur radio. Welcome to a significant depth of the hobby!

CHAPTER 10
Building and Experimenting

You will discover a new horizon as you continue to explore the exciting world of amateur radio, which is one of constructing and experimenting with new things. The hands-on creativity and tinkering that allows you to personalize your pastime to your interests and needs is where the true heart of amateur radio beats. This is where the hobby actually comes from. Suppose you embrace this part of amateur radio. In that case, it is possible to turn yourself from an operator into a creator, an innovator, and even an inventor. This opens up a universe of possibilities. Join me on this adventure as we investigate how constructing and experimenting might further enhance your experience with amateur radio.

Imagine the sense of accomplishment you would feel if you could communicate with someone using an antenna that you had constructed yourself or the excitement you would feel if you used a radio that you had assembled or modified. The profound and indescribable thrill of building something that connects you to the world is what this is all about; it is not just about saving money or customizing your setup. You will gain a deeper understanding and appreciation of the science and art of radio with each project you tackle, whether it be the construction of a straightforward wire antenna, the assembly of a QRP (low power) kit, or the design of intricate digital interfaces.

When you start your first construction job, it can be both thrilling and intimidating at the same time. Start with less challenging, more doable, and appropriate projects for your current level of expertise. When getting started, something as straightforward as a dipole antenna or a basic Morse code practice oscillator can be a fantastic choice. These projects briefly introduce electronic building and design concepts, which require only a few components to execute. With innumerable manuals, movies, and forums where other hams share their ideas and advice, the internet is a treasure trove of resources that may be found on the internet.

The same may be said regarding the amateur radio experimenter: skilled craftsmen are nothing without their tools. A basic set of electronic tools, which includes a multimeter, wire strippers, pliers, and a soldering iron, can help you create a wide variety of projects. As your ambition increases, you may purchase more instruments, such as oscilloscopes, signal generators, and antenna analyzers, for your workshop. These tools will allow you to expand the scope of your constructions and level of sophistication.

Failure is not merely a possibility in development and experimentation; it is an essential component of learning. Only some projects will work out how it was meant on the first attempt, which is acceptable. An error is a lesson disguised as a mistake, providing insights into what works, what does not work, and the reasons behind these differences. The most important thing is to approach each endeavor with patience and an open mind, ready to resolve any issues, make any necessary adjustments, and start again. It is important to remember that the ashes of unsuccessful experiments have been the source of some of the most brilliant ideas in amateur radio.

When it comes to creating and experimenting, the community is one of the most attractive features of amateur radio, and this is especially true in the former. One can find a wealth of inspiration, support, and opportunities for collaboration in local groups, online forums, and maker spaces. To gain knowledge from the experiences of others, you should not be afraid to seek advice, discuss your accomplishments (and failures), and share your own experiences. It is also possible to have a sense of camaraderie and shared accomplishment by participating in collaborative projects, build-a-thons, or courses that focus on constructing kits.

As your self-assurance and skill level increase, the difficulty level of your projects will also increase. Building your own transceiver, experimenting with software-defined radio, or developing antennas optimized for tough operating conditions are all kinds of challenges you can find yourself drawn to. In addition to expanding your technical capabilities, advanced projects contribute to the larger body of knowledge amateur radio enthusiasts hold. By disseminating your discoveries to the community, new means of communication or advancements in radio technology could be made possible.

While creating and experimenting are exciting activities, they also involve the responsibility of doing so safely. When working with electricity, chemicals, or at heights, it is especially important to remind yourself of the potential dangers that may result. You must educate yourself on the appropriate safety standards and always take your time when it comes to safeguarding yourself and others. One definition of a successful experiment is risk-free.

Constructing and experimenting in amateur radio is ultimately a learning process that continues throughout one's life. Your understanding of electronics, radio theory, and the physical laws that govern the airways will expand with each project, regardless of whether or not it is successful. Your interests, the obstacles you face, and the ever-evolving frontiers of technology significantly shape your course.

Remember that you are immersing yourself in the very core of amateur radio as you embark on creating and testing. This is not merely a pastime but an adventure filled with discoveries, creative endeavors, and innovative ideas. Each project presents an opportunity to strengthen your connection to the vast, invisible world of radio waves surrounding us, and your workbench transforms into a launchpad for that journey.

We would like to take this opportunity to welcome you to this exciting new chapter in your amateur radio adventure, where the only limitations are those imposed by your creativity and resolve. In the solder smoke and buzz of equipment, you can discover a great sense of success and a deeper grasp of the magic that is amateur radio. Other than the problems and solutions, you will find these things here.

DIY Projects for Ham Radio

The act of beginning do-it-yourself projects within the amateur radio field is like opening a treasure box that contains infinite possibilities and adventures. Every project offers a voyage into the heart of what makes ham radio such a profoundly enjoyable hobby, regardless of whether it is the construction of a straightforward antenna or the assembly of a complicated transceiver. When you make a solder connection, cut a wire, or test a circuit, you are not only constructing a piece of equipment but also incorporating your own being into the fabric of amateur radio. We are going to investigate some do-it-yourself projects that have the potential to enhance your experience, broaden your knowledge, and strengthen your connection to this fascinating world.

Making a dipole antenna could be your first venture into do-it-yourself projects. This will allow you to begin with something fundamental and vital. This undertaking does not call for a comprehensive inventory of materials or a complex arrangement of components. To create an antenna that may open up the airwaves for communication that is both clear and effective, all you need is some wire, some insulators, a coaxial cable, and a few connectors. In addition to providing you with the joy of collecting signals on something you have constructed with your hands, the process teaches you the fundamentals of antenna design. By tuning your antenna to the frequency band of your choice, you will understand the fundamental concepts of resonance, impedance, and SWR (Standing Wave Ratio), which are fundamental ideas that any ham operator needs to be familiar with.

A one-of-a-kind challenge and opportunity await those who venture into the realm of QRP, which stands for low-power operation. When you put together a QRP kit, you not only become familiar with the complexities of radio circuits but also gain an appreciation for the efficiency and expertise required to create contacts with a low amount of power. Through the completion of this project, you will improve your soldering skills and gain a deeper grasp of radio electronics. There is no feeling quite like the rush you get when you make that first touch on a rig you have put together. A demonstration of the power that may be achieved by knowledge, patience, and precision.

For the ham operator who enjoys working out in the field, constructing a portable power supply becomes a vital job. Incorporating batteries, solar panels, or even hand-crank generators into a small and efficient power solution provides assistance for your portable activities and educates you about power management and renewable energy sources. Because of this project, you can run your equipment in remote areas or during emergencies where standard power sources are unavailable. This project highlights the significance of sustainability in the hobby.

Your ability to explore digital modes can be substantially improved by constructing your own digital interface, which is becoming increasingly popular in amateur radio as digital modes continue to gain popularity. By completing this project, you can transmit digital signals over the airwaves by constructing a bridge between your computer and your radio transmission device. This represents exploring digital protocols, software setup, and sound card technology. Through participation in this project, you will not only be able to expand your operational capabilities, but you will also gain access to modes such as FT8, PSK31, and RTTY, each of which has its own set of diverse opportunities for international communication.

For efficient and risk-free radio operations, you must comprehend and effectively manage the standing wave ratio (SWR) of your antenna system. Constructing your SWR meter is not merely a practical project; rather, it is an in-depth exploration of the fundamental concepts that guarantee the transmission of your signals effectively without causing any damage to your apparatus. By completing this project, the notion of SWR will be made more understandable, and you will be provided with a tool that is important for tuning antennas and identifying transmission problems. Your station's capabilities will be improved due to this combination of electrical engineering engineering and practical radio operation.

Crafting a practice oscillator is a fascinating project that blends simplicity with significant learning benefits. It is ideal for individuals interested in the ageless art of Morse code. With the help of this device, you will be able to improve your Continuous Wave (CW) skills by offering a platform for practice that is both tactile and aural. You will understand the foundations of electronic circuits by constructing them from essential components like resistors, capacitors, a speaker, and a key. In addition, this procedure will prepare you to excel in one of the most revered means of communication in ham radio technology.

Participating in radio direction finding, sometimes known as "fox hunting," is an exciting component of amateur radio that combines the pleasure of the hunt with the technical expertise required. The process of constructing a fox hunt transmitter requires you to develop a gadget that sends out a signal that can be followed by other people. This puts your talents in both creating and transmitting to the test. This project is not only about soldering components; rather, it is about contributing to a collaborative game that improves listening, locating directions, and working together abilities among members of your local amateur radio community.

When one enters the realm of SDR, a universe of exploration becomes available to them. Not only are you putting together hardware when you put together a static digital receiver (SDR) configuration, but you are also opening the door to cutting-edge radio technologies. Through this project, you can receive and decode various signals, providing a comprehensive illustration of the radio spectrum. At the same time that it provides insights into signal processing, digital communications, and the future of amateur radio, it is an investigation of the intersection between traditional radio and the digital world.

Participation in each do-it-yourself project brings you one step closer to entering amateur radio's huge and fascinating world. Not only are these projects duties that need to be finished, but they are also experiences that need to be lived, and each one presents its own distinct mix of difficulties, opportunities, and pleasures. Remember that you are a part of a continuum of amateur radio operators who have, for centuries, pushed the bounds of communication. This is something to remember as you solder, construct, and experiment.

Construction and experimentation are the two most important aspects of amateur radio. It is about putting what you have learned into practice and making it your own by applying it practically. The charm of do-it-yourself (DIY) in amateur radio is unrivaled, not only because of the simple thrill of a wire antenna catching its first signal but also because of the complicated gratification of a home-built transceiver making its first contact.

Antenna Desing and Construction

In the course of your exploration of the realm of amateur radio, you will quickly come to realize that the process of designing and constructing antennas is not merely a task; rather, it is an art form, a science, and a journey of discovery all rolled into one wonderful experience. Between your radio and the great expanse of the airwaves, the antenna serves as the bridge that connects the two. This is the key to unlocking clear and effective communication skills, whether you are trying to communicate with a neighbor or someone on another continent. Let's go on this adventure together, learning the foundations of antenna design and construction to ensure you are well-equipped to make antennas that enhance your experience with amateur radio.

To construct an effective antenna, it is crucial to have a solid grasp of the fundamental concepts that form the basis of antenna design. Whether transmitting or receiving radio waves, the primary function of an antenna is to radiate them efficiently. A key concept to understand is resonance, which occurs when the length of the antenna matches the wavelength of the frequency being used. This harmony between the antenna length and the wavelength enables efficient energy transfer, thereby maximizing the range and clarity of your signal.

The dipole antenna, one of the most fundamental yet diverse antenna designs, acts as a great starting point for any building projects you may be working on. The dipole, a core model for understanding how antennas function, comprises two conducting components of equal length that are arranged end to end. Constructing a dipole will teach you fundamental skills such as measuring and cutting wire, soldering connections, and adjusting for resonance. The feeling of accomplishment that comes from making your first successful contact on an antenna that you constructed yourself is quite satisfying and demonstrates your ability and inventiveness.

Your self-assurance will increase, and your interest in other antenna designs will also increase. Every form of antenna has its own benefits and difficulties, ranging from the vertical antenna's ability to be small and convenient to the Yagi antenna's ability to provide directional coverage. Experimenting with various designs allows you to customize your setup to meet your particular requirements and the conditions in which you operate. The realm of antenna design is where you will find the answer to your antenna needs, whether searching for an antenna that can perform effectively in confined spaces or zero in on far-away signals.

The concept of impedance matching refers to ensuring that the impedance of your antenna is identical to that of your radio and feedline. This is an essential component of antenna building. The Standing Wave Ratio (SWR) measures a phenomenon that prevents radio waves from reflecting back towards your transmitter. This harmony prevents this occurrence from occurring. Learning how to test and adjust your antennas' standing wave ratio (SWR) is essential for ensuring that your equipment is protected from damage and that your operation is efficient. By constructing and utilizing devices such as antenna tuners and SWR meters, you will acquire a more profound comprehension of these ideas, improving your ability to construct and operate.

The success of your antenna projects may be influenced by the materials you select and the equipment you employ. Even though many antennas may be constructed using straightforward and easily accessible materials like wire, PVC pipes, and coaxial cables, the quality of these materials can impact the antenna's performance and durability. Similarly, outfitting oneself with the appropriate tools, such as a dependable soldering iron, wire cutters, and possibly even a drill, will make the construction process easier and more pleasurable.

Regarding antenna building, paying close attention to safety is necessary, particularly when the work entails climbing or working at heights. Whether you put an antenna on a roof, a pole, or a tree, you should always put your health and safety first and equip yourself with the appropriate safety gear and methods. Additionally, pay attention to the placement of the antenna near power lines and any other potential dangers.

One of the most attractive features of amateur radio is the network of other enthusiasts who are eager to share their expertise and experiences with one another. Participating in this community through local clubs, internet forums, or hamfests can provide extremely helpful insights into the design and construction of antennas. You can speed up your development by gaining knowledge from the achievements and failures of other people. This can motivate you to take on new challenges and push the limits of what you previously believed achievable.

The opportunity for invention and experimentation is the essence of antenna design and construction, which is the ultimate beauty of these processes. An activity that thrives on innovation is amateur radio, a hobby. As you construct antennas, you not only add to your understanding of the pastime and the enjoyment you derive from it but also contribute to the overall body of information held by the community of amateur radio operators. Your contributions are valuable, regardless of whether you are going to improve the performance of a traditional design or you are going to venture into uncharted territory with a novel concept.

It is important to remember that every project, regardless of how big or tiny it may be, represents a step forward in your amateur radio adventure as you reflect on your trip through antenna design and building. In your continued investigation of the airwaves, each antenna you construct represents a story of learning, a difficulty you have conquered, and a milestone you have reached.

Take advantage of the fact that constructing antennas and experimenting with them is a crucial and enriching component of your experience with amateur radio. Your talents will be tested, your knowledge will be expanded, and eventually, you will feel a deeper connection to the world around you as you embark on this adventure. We would like to take this opportunity to welcome you to the fascinating world of antenna design and construction, where your imagination and curiosity will lead you to new directions in the realm of communication.

Homebrewing Equipment and Modifications

In amateur radio, entering the realm of homebrewing equipment and modifications is analogous to entering a workshop where innovation and tradition coexist, and creativity has no limits. This adventure you are about to embark upon is not just about playing with electronics; rather, it is about embracing the essence of amateur radio, which is the spirit of exploration and the delight of personal achievement. As we work together to unravel this complicated tapestry, we will ensure that you approach this adventure with self-assurance, excitement, and the certainty that each step ahead will deepen your experience of amateur radio.

Homebrewing, in the context of amateur radio, refers to the process by which amateurs create or modify radio equipment of their own accord. This is done not simply for the sake of construction but also to tailor, improve, or even establish new forms of communication. To accomplish this, you might have to design and build your own transceiver, create a one-of-a-kind antenna tuner, or alter current equipment to improve its functionality or add new capabilities. This process demonstrates the operator's inventiveness, technical expertise, and enthusiasm for the activity they are participating in.

At its foundation, homebrewing is about satisfying demands that can't be addressed by off-the-shelf equipment. This could be the challenge of accomplishing something with limited means, the need for a customized solution to a particular problem, or the pure pleasure of creating something new. The more projects you take on, the more personal your station will become, and the more it will reflect your individuality. In addition, the information and experience you acquire via these projects will help you develop a deeper understanding of radio theory and electronics, both of which are extremely significant assets in amateur radio.

You should begin with the fundamentals before moving on to more complicated projects. You should get familiar with essential electronic components such as resistors, capacitors, transistors, and integrated circuits, and you should also grasp the roles these components play inside a circuit. Constructing a straightforward Morse code practice oscillator, an audio filter, or a QRP (low power) antenna tuner are good projects that can serve as ideal beginning places. Not only will you improve your soldering and circuit design skills by completing these projects, but you will also discover the sense of self-assurance that comes with successfully bringing a creation to life.

For homebrewing to be successful, having a well-equipped workstation is essential. Some of the essential tools are a soldering iron, wire strippers, a digital multimeter, and a variable power supply. As you advance, you may acquire further specialist apparatus such as an oscilloscope, a signal generator, and an antenna analyzer. These tools help with developing and testing, as well as debugging and improving the projects you are working on.

A wealth of information and ideas is available to homebrew enthusiasts within the community of amateur radio transmitters. Project ideas, schematics, and advice from other amateur radio enthusiasts may be found in online forums, local clubs, and magazines for amateur radio. Participating in community activities stimulates your creative process and provides you with support and encouragement. Additionally, working together on projects or participating in homebrew competitions can be extremely fulfilling, as they provide opportunities to gain new perspectives and face new problems.

Keeping detailed records of your projects is as essential as the construction process itself. When you keep extensive records of your designs, adjustments, and the logic behind specific choices, you provide a helpful reference for future projects and make it easier to share your inventions with the community. Other hams can imitate your achievements or build upon your concepts with detailed documentation, which can help develop a culture of collaboration and continual progress over time.

Experimentation is not without its dangers, especially when it involves experimenting with electricity. Always make safety your top priority; always wear the appropriate protective gear, always be aware of high voltages, and never bypass the safety mechanisms on any equipment. Understanding the potential dangers and how to protect yourself from them is essential if you want your journey into homebrewing to be fruitful and risk-free.

The process of homebrewing is full of difficulties. Inevitably, you will experience failures, such as circuits that do not function as anticipated, unsuccessful design designs, or alterations that do not result in the desired improvements. Even though they are annoying, these moments are filled with learning opportunities. Each problem you solve, and every error you fix contributes to the breadth and depth of your knowledge and competence.

The genuine satisfaction of homebrewing is the realization that your creations have been brought to life, that they have served their intended purpose, or even that they have beyond your expectations. The satisfaction that may be obtained from these successes is unrivaled, regardless of whether it is a home-built transceiver establishing its first contact or a customized antenna tuner precisely matching impedances. It physically represents your inventiveness, skilled craftsmanship, and unwavering commitment to the pastime.

Continue to learn more about homebrewing. You will discover that the possibilities are as boundless as your imagination allows. Depending on the circumstances, today's project could be a straightforward antenna swap, while tomorrow's project could be a complex SDR setup or a bespoke digital mode interface. With each component soldered, circuit built, and change executed, you are not only constructing equipment; you are also shaping the future of your adventure via amateur radio.

Homebrewing equipment and modifications are fundamental components of amateur radio, and they are a key feature that is not only tough but also quite gratifying. Embodied with limitless spirit, the journey of amateur radio is characterized by ongoing learning, creativity, and personal expression through interest. Every project in amateur radio is a step towards mastery and shows dedication to changing communication environments.

CHAPTER 11

Continuing Your Ham Radio Journey

When you first started on your trip through amateur radio, you went through the basics, tried your hand at DXing, experimented with home-brewed equipment, and prepared yourself to communicate in an emergency. The route of amateur radio, on the other hand, never ends; it is a never-ending voyage of discovery, progress, and social interaction. You should know that the journey is ongoing as you stand at this crossroads and stare towards the horizon. Together, we will investigate how you can continue to enhance your adventure with ham radio and embrace the countless opportunities that are still to come.

The landscape of amateur radio is as large as it is varied. It encompasses the technical aspects of radio operation, the rich history, the ever-changing regulatory environment, and the cutting-edge technological breakthroughs used to shape the hobby. You can improve both your comprehension of amateur radio and your appreciation for it by devoting some time to expanding your knowledge through books, online courses, and attendance at community lectures. A fruitful ground for inquiry and learning can be found in digital communication, antenna theory, and radio wave propagation fields.

As your interest in the pastime increases, so will the size of your station. By upgrading your equipment, you can dramatically improve your ability to make connections, explore new modes, and participate in contests. This can be accomplished by purchasing new gear or tweaking your setup. Suppose you want to enhance the capabilities of your station. Consider purchasing a more powerful transceiver, a mobile antenna system, or advanced software. The implementation of each change not only increases the efficiency of your operations but paves the way for new opportunities for exploration and inquiry.

The strength of the community that supports amateur radio, a worldwide network of enthusiasts united by a common interest in communication, is what makes the hobby so successful. By participating in this community on a deeper level, whether through local clubs, online forums, or international groups, one can receive support, inspiration, companionship, and other benefits. As a method to give back to the amateur radio community and deepen the relationships that exist within it, one can volunteer for club roles, organize events, or mentor new hams.

It is still a captivating component of the adventure of amateur radio to strive for prizes and certificates, as these provide motivating and challenging goals. A few less popular recognitions need contacts to be made using satellite, digital modes, or QRP operations. These are in addition to the standard DXCC or WAS honors. Your commitment and the abilities you've polished over many hours of operation are reflected in each award, which signifies a milestone that has been reached and a journey that has been traveled.

Each band and mode on the radio spectrum presents obstacles and opportunities, making the radio spectrum a paradise of possibilities. Rekindling your interest and enthusiasm for the sport can be accomplished by venturing into bands that have not been explored before or by playing with modes such as moonbounce (EME), meteor scatter, or even amateur television (ATV). These investigations not only push the limits of what you believed was possible with amateur radio, but they also contribute to the body of knowledge and invention associated with the hobby.

To putTo put your abilities to the test, enhance your productivity, and engage in a spirit of friendly competition, contests, and special event stations offer dynamic platforms. When you participate in these activities, whether as a lone operator or as a club member, you are challenged to operate under pressure, successfully manage resources, and reach for greatness. Each competition is an opportunity to acquire new knowledge, develop one's skills, and rejoice in the pleasures of amateur radio within a worldwide community.

When assisting, your talents as an amateur radio operator offer extraordinary potential. You will be better prepared to assist your community during times of crisis if you increase your involvement in emergency communication through groups such as ARES or RACES, but you will also be able to demonstrate the fundamental idea of amateur radio service. By understanding that your enthusiasm for radio may have a discernible and constructive effect on other people's lives, this facet of the pastime provides a tremendous sense of fulfillment.

The need to remain educated on changes in laws, best practices, and new technologies must be supported because technological landscapes and regulations are always shifting. Maintaining compliance with regulatory regulations ensures your continuing enjoyment of the pastime without interruption while keeping you aware of technical changes that might inspire new projects, modes of operation, and methods of communication.

To maintain progress and happiness in the pastime, it is vital to take the time to reflect on your path, celebrate your accomplishments, and establish new goals for yourself. Creating goals for yourself can give you direction and purpose, driving your passion and curiosity for amateur radio. Whether it's mastering a new mode, attaining a tough award, or simply making more contacts than the previous year, creating goals.

Suppose you want to continue your adventure in amateur radio. In that case, the most important thing you can do is commit to learning throughout your life. Whether it's a new technical skill, a piece of history, or a greater awareness of the global society you're a part of, every day brings an opportunity to learn something besides what you already know. Every frequency you tune, every contact you make, and every project you take on contributes to the enrichment of your life and connects you to a world beyond your own. Amateur radio is not simply a hobby but a journey of continual discovery.

Remember that the horizon extends through amateur radio as you continue on your adventure. As you move forward, you will be weaving the threads of learning, building, exploring, and serving into the fabric of your voyage. The beauty of this voyage is not in the destination itself but rather in the experiences, friendships, and discoveries made along the road at various points along the path. The airways are alive with the potential of fresh knowledge, connections, and the unending joy of communication. We would like to take this opportunity to welcome you to the lifelong adventure that is amateur radio.

Advancing your Skills and Knowledge

Since you have decided to embark on the path of amateur radio, you have accepted a world in which the airways connect you not only to other enthusiasts but also to a never-ending chance for personal development and career advancement. To advance your skills and knowledge in this dynamic activity, it is not enough to just chase after faraway signals or accumulate equipment; rather, it is necessary to cultivate a profound and enduring love for making discoveries and achieving mastery. Join me as we go along this route, and together, we will investigate how you might enhance your trip through amateur radio by enhancing your understanding and polishing your talents.

Your insatiable curiosity is the driving force behind your path through amateur radio; it is an unquenchable thirst to learn, participate, and become more involved in the sport. Encourage this sense of wonder. Allow it to direct you to new bands, modes, and radio features you have yet to investigate. Whether you are interested in delving into the mysteries of digital modes, the complexities of antenna construction, or the subtleties of radio wave propagation, your curiosity is the compass that guides you to a road that leads to deeper knowledge and competence.

Education serves as the cornerstone upon which advancement in amateur radio is constructed. Devote some of your time to studying to ensure that you pass the license exams and grasp the fundamentals of what makes your radio function properly. Learn to think critically about radio theory, electronics, and communications by looking for materials such as books, online courses, webinars, and club lectures that will push you to think critically about these topics. Not only is the acquisition of knowledge necessary for passing examinations, but it also enhances your experience and makes it possible for you to participate in the pastime more profoundly.

The community of amateur radio includes a large number of people who have a wealth of expertise and knowledge in their respective fields. Involve yourself with these guides. Gain knowledge from both their achievements and their setbacks. You can receive direction from a mentor who can motivate you to try new things and provide insights that cannot be found in books or classes. It can expedite your learning and open doors to new opportunities by cultivating relationships with more experienced operators. This can be accomplished through your local radio club, internet forums, or hamfests.

When you use equipment you have constructed or modified yourself, you experience a level of enjoyment unlike any other. Participate in initiatives that will test your knowledge and force you to learn new abilities regularly. Each project, whether it is the construction of a straightforward antenna, the assembly of a kit transceiver, or the design of a complex digital interface, helps you develop your technical expertise and provides a hands-on understanding of the practical side of amateur radio.

Contests and special event stations offer opportunities to polish your operating abilities and develop effective communication methods in a variety of settings, which are unmatched in the industry. Every competition serves as a learning laboratory, providing obstacles that put your skills to the test and encourage you to better yourself. Not only can participation help you improve your abilities, it also helps you become more self-assured and connects you with a larger community of operators who share your enthusiasm for the hobby.

When you pursue higher levels of licensure, you are not only committing to gaining extra operating rights; you are also making a commitment to advancing your knowledge and understanding of amateur radio. You will be required to acquire knowledge of more difficult ideas as you progress up the licensure ladder. These concepts range from advanced electronics to sophisticated communication strategies. Achieving a higher-class license is a significant accomplishment that demonstrates your commitment to the hobby and enables you to explore new facets of the world of amateur radio.

Not only does sharing your expertise and experiences with others contribute to the growth of the amateur radio community, but it also helps you strengthen your own learning. Give a talk at your neighborhood club, teach a lesson on obtaining a license, or write articles for magazines geared toward amateur radio. Teaching and sharing forces you to arrange your knowledge and comprehend it as well as possible so that you can explain it to other people, ultimately improving your abilities and comprehension.

A constant stream of new technology, rules, and operational procedures is being introduced into amateur radio, a field that is constantly undergoing change. To stay informed about the latest changes in amateur radio, there are a few things you can do. Subscribing to amateur radio periodicals, participating in online forums, and attending hamfests and conferences are ways to keep current. By doing this, you can ensure that your knowledge remains relevant and your abilities remain useful.

It is important to schedule regular time to evaluate your progress, acknowledge your accomplishments, and plan for the future. Through reflection, you can recognize the extent to which you have progressed and pinpoint the areas in which you wish to develop further. Setting clear and attainable goals can keep your trip through amateur radio focused and purposeful. This will ensure that you continue to increase your horizons and move forward in your journey.

Regarding it, embracing lifelong learning is the most important thing you can do to advance your abilities and knowledge in amateur radio. From the technical to the operational, from the historical to the futuristic, the hobby provides a limitless curriculum of things to master. The disciplines range from the technical to the operational. Approach each day as an opportunity to learn something new, experiment, ask questions, and push the boundaries of your understanding. This is how you should approach each day.

Keep in mind that advancement in amateur radio is not only about collecting certifications or accolades; rather, it is about the personal gratification that comes from mastering difficult problems, contributing to the community, and connecting with people and ideas that inspire you. Keep this in mind as you continue on your journey. Your journey toward becoming an amateur radio operator demonstrates your inquisitiveness, commitment, and willingness to venture into the unknown.

Upgrading Your License

Upgrading your amateur radio license is one of the most important parts of your ham radio adventure. This choice leads to deeper investigations into the field of radio communication, wider perspectives, and more intricate discussions. Remember that getting an upgraded license is more than access to more frequencies when considering this next step. It's about pushing yourself, learning new things, and realizing the full potential of amateur radio.

Gaining access to further rooms in the enormous amateur radio estate is akin to getting an upgraded license. With access to extra frequencies offered by each new license class, you can now explore bands that were previously unreachable. These new frequencies may allow you to experiment with more bandwidth-intensive communication methods or clearer channels. Envision the excitement of successfully reaching out to a band that was merely a line on a chart but is now a reality because of your perseverance and hard work.

At its foundation, renewing your license is a learning process. You will study in-depth subjects such as sophisticated electronics theory, intricate operational procedures, and in-depth regulatory understanding. This isn't just about passing an exam; it's also about deepening your awareness of the laws and science underlying your favorite pastime. Consider this learning a chance to improve, hone your abilities, and become a more skilled operator and a more knowledgeable part of the amateur radio community rather than as a barrier.

This is not an isolated adventure for you. The mentoring and supportive nature of the amateur radio community is well known. Discover how to take advantage of this information by joining nearby groups, participating in internet discussion boards, or locating an experienced mentor known as "Elmer," who can help you along the way. Many resources are available for study, including books, online courses, and practice tests that mimic real exams. By creating a well-structured study plan that suits your learning style and lifestyle, you may use these tools to set realistic objectives and checkpoints.

Although getting ready for an upgrade can be difficult, remember that each obstacle you overcome brings you closer to mastery. You will feel great accomplishment when you pass your exam because of the hours you spend studying, the effort you put forth to comprehend difficult ideas, and your perseverance in moving forward even when the subject matter appears overwhelming. With a strong sense of purpose and excitement, take on this task, knowing that the benefits will far outweigh the additional rights your updated license will confer.

When exam day finally comes, go in confidently. You've learned a great deal, practiced, and prepared yourself. The test is not a barrier but a turning point in your amateur radio career—a concrete indication of your progress and commitment to the hobby. Rest well the night before the test, drink enough water, and have faith in your study skills. Your diligence and enthusiasm for amateur radio are evident in every accurately answered question.

It's exciting to pass your exam and get your license upgraded. This is an accomplishment in and of itself, but it also merits praise for what it stands for: your dedication to a pastime that unites people from all walks of life, your readiness to take on new challenges, and your commitment to lifelong learning. Revel in the company of your fellow hams, impart your wisdom to those contemplating an upgrade, and pause to realize how far you've come.

Now that you have your updated license, you can access a more varied and rich amateur radio environment. You have fresh tasks to overcome, new modes to try, and new frequencies to investigate. More than that, though, is the fact that you have demonstrated to the community and to yourself your dedication to developing professionally and making a positive impact on the amateur radio community.

Remember that getting an upgraded license marks a new beginning rather than the conclusion of your amateur radio career. It's a call to keep delving deeper and more purposefully into the pastime via experimentation and exploration. Anticipate the new experiences, relationships, and learning opportunities ahead of you. A license upgrade signifies how far you've come and how far you can go on your amateur radio journey, which is a lifetime experience.

Staying Engaged with the Community

Within the dynamic realm of amateur radio, your experience is enhanced by the frequencies you tune into and the people you interact with. Maintaining ties with this community is like creating a beautiful tapestry of shared experiences, wisdom, and friendships. It's about more than just networking; it's about forming connections, picking up knowledge from others, and adding to a culture that values cooperation, creativity, and communication. Let's explore ways that you can continue to be involved in the amateur radio community so that your adventure continues to be alive, rewarding, and enhanced by the global community of hams.

The local amateur radio groups serve as the community's lifeblood. By joining a group, you can meet other aficionados, exchange stories, and discover new facets of your interest. Clubs frequently provide get-togethers, guest speakers, and practical seminars that can increase your knowledge and abilities. Engaging in these activities improves your knowledge and fortifies links within the amateur radio community in your area. Seize the chance to help by offering your knowledge, helping at club functions, or just listening to other hams tell their story.

The internet amateur radio community is thriving in today's interconnected world. Digital ham nets, social media groups, and forums provide places to explore, discuss, and debate every imaginable facet of amateur radio. By connecting you with hams worldwide, these platforms can provide various viewpoints and ideas outside your area. Actively participate while maintaining decorum, and always remember that each call sign represents a person with their unique story and passion for the pastime. Ask questions, share your experiences, and give guidance; each exchange adds a new thread to the global amateur radio network.

Serving the public and responding to emergencies as a volunteer is one of the most rewarding ways to get involved in the community. Amateur radio operators frequently provide a vital communications link for events such as marathons, festivals, and disaster relief operations. Participating in these events as a volunteer shows how important amateur radio is to the larger community and allows you to hone your talents in practical situations. It's a meaningful way of giving back, demonstrating that amateur radio is an important public service and hobby.

For amateur radio enthusiasts, hamfests and conventions are like pilgrimages—a chance to get together, share, educate, and enjoy the hobby. You can meet amateur radio personalities, examine the newest equipment, and participate in lectures and workshops by attending these events, which can be very inspirational. But what enhances your trip is the opportunity to interact in person with people who share your enthusiasm, which goes beyond technology and education. Your relationship with the amateur radio community is strengthened by each handshake, tale shared, and advice offered or taken.

Sharing your experiences, discoveries, and inventions with a larger audience can be accomplished through writing articles or submitting content to blogs, newsletters, or amateur radio magazines. Your contributions enrich the community's knowledge base, whether offering advice for novices, describing a project you've finished, or writing about your experiences using emergency communications. Writing benefits others and enables you to think back on your experiences, strengthening your comprehension and enjoyment of the pastime.

Do you recall the thrill and maybe even the fear you experienced when you started? You now have the chance to mentor newbies to the pastime as an experienced operator. Assisting novice radio operators by providing direction, responding to inquiries, and expressing your excitement can ease their initial difficulties and cultivate a passion for amateur radio. One of the most straightforward methods to support the development and sustainability of the community and make sure the pastime survives for future generations is through mentoring.

Ultimately, the key to remaining active in the amateur radio world is to keep an open mind and a feeling of inquiry. This ever-evolving hobby always develops new concepts, approaches, and technologies. You can ensure that your amateur radio journey continues to be exciting and rewarding by maintaining your curiosity and being open to new experiences. In addition to expanding your knowledge, your willingness to try and learn new things can encourage others in the community to delve deeper and pursue their goals.

Maintaining an active involvement in the amateur radio community means more than just taking part; it also means helping to shape the culture of sharing, education, and camaraderie that characterizes the sport. As you proceed on your amateur radio adventure, remember that your contacts, contributions, and excitement enhance your own experience and the experience of hams worldwide. Greetings from a thriving global community where every interaction, project, and conversation is a thread in amateur radio's ongoing adventure.

Appendix: Glossary of Ham Radio Terms

Entering the realm of amateur radio is like stepping into a technologically advanced, historically rich, and socially diverse group. You will come across a wide range of acronyms and terminology essential to understanding ham radio communication as you navigate this interesting terrain. Gaining proficiency in this distinct language is not the only goal; it's also about strengthening your bond with the pastime and the international society surrounding it. To help you become a proficient member of the amateur radio community, let's examine some of the key ham radio lingo.

Audio Frequency Shift Keying, or AFSK: An audio-tone-based digital signal transmission method. Sending text or data over the airwaves is like transforming digital data into a language that radios can comprehend and pronounce.

Antenna Tuner: An antenna tuner is a gadget that modifies the antenna and radio so they function as harmoniously as possible. While it does not 'tune' the antenna per se, it does optimize the power output by adjusting the impedance perceived by the transmitter.

Beacon: Although a beacon in ham radio isn't the same as a lighthouse, it's significant. Similar to putting out a flare to gauge how far and well your signals are moving, a beacon is a transmission from a known point used for propagation or signal route testing.

CW (Continuous Wave): Morse code is the original text messaging system. It's an alternative to speaking out loud, as you can communicate using short (dots) and lengthy (dashes) signals.

Dipole: Picture a clothesline that transmits and receives radio waves instead of clothes. The dipole is one of the easiest and most efficient antennas you may use or create.

DX: If Ham Radio is a world of adventurers, DX represents far-off places. It represents long-distance communications, where the excitement of getting in touch with someone far away always remains the same.

Elmer: Every hero needs a mentor, and in amateur radio, that mentor is known as an Elmer. They transfer the understanding of the pastime by mentoring, instructing, and inspiring newbies.

FCC: The head of American communications is the Federal Communications Commission (FCC). The FCC ensures that the airwaves are utilized fairly and responsibly by establishing the rules, allocating frequencies, and granting licenses.

Gain: It's about maximizing what you already have, not acquiring anything. "gain" refers to the antenna's ability to focus or direct energy. Your signal travels farther with more gain, but direction is crucial.

Handheld Transceiver: The handheld transceiver, known as the HT (Handie-Talkie), is your radio companion that enables you to converse while on the go. Your HT is there for you in an emergency or at a hamfest on a hilltop.

Ionosphere: The radio waves' magic carpet is not just a stratum of Earth's atmosphere. Thanks to its ability to reflect or bend signals back to Earth, signals can pass beyond the horizon.

Q Codes: A collection of acronyms that begin with "Q," utilized to facilitate effective communication. QSO, QTH, or QRP? These are the amateur radio equivalent of shorthand notes, which speed up and simplify intricate conversations.

Repeater: Consider it similar to a radio relay service. Your transmission range increases when it receives your signal and rebroadcasts it at a higher power and/or from a better position.

RF (Radio Frequency): These waves, which transmit data, voice, and Morse code via the air, are the lifeblood of amateur radio. The unseen threads that bind hams worldwide are made by radio frequency (RF).

Single Sideband, or SSB: Using only one portion of the modulated signal, SSB, an efficient relative of AM (amplitude modulation), allows you to talk farther on less power. It's similar to conversing in a noisy environment but with greater volume and clarity.

Standing Wave Ratio (SWR): This is an indicator of how well your radio and antenna are working together. A high SWR is like shouting into a pillow, but an ideal SWR is when all your force radiates out into the world.

UTC: The global ham radio clock is coordinated Universal Time, or UTC. UTC ensures everyone is in sync and arrives on time for nets, contests, and skeds, regardless of where they are.

When you first start using ham radio, this glossary will serve as your guide, helping you understand the hobby's fundamental terms and ideas. Gaining proficiency in a new term doesn't simply help you learn new phrases; it also makes you a more well-rounded amateur radio operator equipped to interact, discover, and connect with other enthusiasts worldwide. Greetings from the fascinating and dynamic world of amateur radio, where new terms mean exciting discoveries every day.

www.ingramcontent.com/pod-product-compliance
Lightning Source LLC
Chambersburg PA
CBHW062227220526
45471CB00009B/3375